T0271852

Electric and Electronic Circuit Simulation using TINA-TI®

RIVER PUBLISHERS SERIES IN CIRCUITS AND SYSTEMS

Series Editors:

MASSIMO ALIOTO
National University of Singapore
Singapore

KOFI MAKINWA
Delft University of Technology
The Netherlands

DENNIS SYLVESTER
University of Michigan
USA

The "River Publishers Series in Circuits and Systems" is a series of comprehensive academic and professional books which focus on theory and applications of Circuit and Systems. This includes analog and digital integrated circuits, memory technologies, system-on-chip and processor design. The series also includes books on electronic design automation and design methodology, as well as computer aided design tools.

Books published in the series include research monographs, edited volumes, handbooks and textbooks. The books provide professionals, researchers, educators, and advanced students in the field with an invaluable insight into the latest research and developments.

Topics covered in the series include, but are by no means restricted to the following:

- Analog Integrated Circuits
- Digital Integrated Circuits
- Data Converters
- Processor Architecures
- System-on-Chip
- Memory Design
- Electronic Design Automation

For a list of other books in this series, visit www.riverpublishers.com

Electric and Electronic Circuit Simulation using TINA-TI®

Farzin Asadi

Maltepe University, Turkey

LONDON AND NEW YORK

Published 2022 by River Publishers
River Publishers
Alsbjergvej 10, 9260 Gistrup, Denmark
www.riverpublishers.com

Distributed exclusively by Routledge
4 Park Square, Milton Park, Abingdon, Oxon OX14 4RN
605 Third Avenue, New York, NY 10017, USA

Electric and Electronic Circuit Simulation using TINA-TI® / by Farzin Asadi.

Routledge is an imprint of the Taylor & Francis Group, an informa business

ISBN 978-87-7022-686-8 (print)
ISBN 978-10-0077-346-0 (online)
ISBN 978-1-003-33279-4 (ebook master)

Dedicated to my lovely brother and sisters Farzad, Farnaz and Farzaneh.

Contents

Preface   xi

List of Tables   xiii

1 **Simulation of Electric Circuits with TINA-TI**®   **1**
   1.1 Introduction   1
   1.2 Installation of TINA-TI   1
   1.3 Version of Software   7
   1.4 Example 1: Simple Resistive Voltage Divider   7
   1.5 Example 2: Volt Meter and Ampere Meter Blocks   34
   1.6 Example 3: Open Circuit and Current Arrow Blocks   39
   1.7 Example 4: RLC Circuit with Non-zero Initial Condition . .   41
   1.8 Example 5: Exporting the Obtained
        Waveforms as a Graphical File   62
   1.9 Example 6: Exporting the Obtained Waveforms
        as Text File   63
   1.10 Example 7: RLC Circuit with Zero Initial Condition   66
   1.11 Example 8: Initial Condition Blocks   72
   1.12 Example 9: Importing the TINA-TI Analysis Result into
        MATLAB®   76
   1.13 Example 10: Measurement of Phase Difference   84
   1.14 Example 11: Power Meter Block   89
   1.15 Example 12: Drawing the Instantaneous Power
        Waveform (I)   93
   1.16 Example 13: Drawing the Instantaneous Power
        Waveform (II)   99
   1.17 Example 14: Ohm Meter block (I)   114
   1.18 Example 15: Ohm Meter Block (II)   117
   1.19 Example 16: Thevenin Equivalent Circuit   121
   1.20 Example 17: Measurement of Thevenin Resistance   129
   1.21 Example 18: Current Controlled Voltage Source   131

1.22 Example 19: Voltage Controlled Current Sources Block . . .   134
1.23 Example 20: Switch Block   138
1.24 Example 21: Three Phase Source   143
1.25 Example 22: Jumper Block   145
1.26 Example 23: Coupled Inductors   151
1.27 Example 24: Transformer   159
1.28 Example 25: Unit Impulse Response of Electric Circuits . .   165
1.29 Example 26: Unit Step Response of Circuits   170
1.30 Example 27: Frequency Response of Electric Circuits (I) . .   174
1.31 Example 28: Frequency Response of Electric Circuits (II) . .   180
1.32 Example 29: Input Impedance of Electric Circuits   185
1.33 Example 30: Drawing the Input Impedance of Electric
        Circuits   189
1.34 Example 31: Phasor Analysis   192
1.35 Example 32: Parameter Sweep Analysis   197
1.36 Exercises   207
        References   210

2   **Simulation of Electronic Circuits with TINA-TI®**   **211**
2.1   Introduction   211
2.2   Example 1: Half Wave Rectifier   211
2.3   Example 2: Measurement of Average and RMS Values
        of Waveforms   216
2.4   Example 3: Harmonic Content of Waveforms   222
2.5   Example 4: Fourier Analysis   228
2.6   Example 5: Converting a Waveform into Sound   231
2.7   Example 6: DC Transfer Characteristics (I)   231
2.8   Example 7: DC Transfer Characteristics (II)   234
2.9   Example 8: DC Transfer Characteristics (III)   237
2.10 Example 9: Temperature Analysis   240
2.11 Example 10: Addition of SPICE Models to TINA-TI® . . .   244
2.12 Example 11: Switching Behavior of Diodes   250
2.13 Example 12: Small Signal AC Resistance of Diodes   257
2.14 Example 13: Full Wave Rectifier (I)   262
2.15 Example 14: Full Wave Rectifier (II)   272
2.16 Example 15: Controlled Rectifier   284
2.17 Example 16: Measurement of Operating Point of Common
        Emitter Amplifier   305

2.18 Example 17: Measurement of Voltage Gain for Common
       Emitter Amplifier 313
2.19 Example 18: Total Harmonic Distortion (THD) of Common
       Emitter 317
2.20 Example 19: THD of Common Emitter Amplifier (II) 322
2.21 Example 20: Frequency Response of Common Emitter
       Amplifier (I) 325
2.22 Example 21: Frequency Response of Common Emitter
       Amplifier (II) 332
2.23 Example 22: Input Impedance of Common Emitter
       Amplifier 336
2.24 Example 23: Output Impedance of Common Emitter
       Amplifier 351
2.25 Example 24: Measurement of Input/Output Impedance with
       Ohm Meter Block 352
2.26 Example 25: Modeling a Custom Bipolar Transistor 355
2.27 Example 26: Modeling a Custom Field Effect Transistor . . 357
2.28 Example 27: Generating the List of Circuit Components . . 358
2.29 Example 28: Non Inverting op amp Amplifier 359
2.30 Example 29: Stability of op amp Amplifiers 371
2.31 Example 30: Measurement of DC Operating Point 382
2.32 Example 31: Measurement of Common Mode Rejection
       Ratio (CMRR) 386
2.33 Example 32: Astable Oscillator 397
2.34 Example 33: Buck Converter 403
2.35 Example 34: Operating Mode of Converter 417
2.36 Example 35: Generating a Pulse with Desired Duty Cycle . . 422
2.37 Exercises 436
       References 439

Index   439

About the Author   441

Preface

A computer simulation is an attempt to model a real-life or hypothetical situation on a computer so that it can be studied to see how the system works. By changing variables in the simulation, predictions may be made about the behavior of the system. So, computer simulation is a tool to virtually investigate the behavior of the system under study.

Computer simulation has many applications in science, engineering, education and even in entertainment. For instance, pilots use computer simulations to practice what they learned without any danger and loss of life.

A circuit simulator is a computer program which permits us to see the circuit behavior, i.e. circuit voltages and currents, without making it. Use of circuit simulator is a cheap, efficient and safe way to study the behavior of circuits. A circuit simulator even saves your time and energy. It permits you to test your ideas before you go wasting all that time building it with a breadboard or hardware, just to find out it doesn't really work.

Toolkit for Interactive Network Analysis (TINA®) is a powerful yet affordable SPICE based circuit simulation and PCB design software package for analyzing, designing, and real time testing of analog, digital, VHDL, MCU, and mixed electronic circuits and their PCB layouts. This software was created by DesignSoft. TINA-TI is a spinoff software program that was designed by Texas Instruments (TI®) in cooperation with DesignSoft which incorporates a library of pre-made TI components to for the user to utilize in their designs.

This book shows how a circuit can be analyzed in TINA-TI® environment. Students of engineering (for instance, electrical, biomedical, mechatronics and robotic to name a few), engineers who work in industry and anyone who want to learn the art of circuit simulation with TINA-TI can benefit from this book.

There are three very good "every day" uses for TINA-TI during your studies: First, it is a very handy tool for verifying lab results. That is, you can recreate a lab circuit, simulate it, and compare the simulation to both your theoretical calculations and lab measurements. Second, it is a handy tool for checking homework if you get stuck on a problem. Third, it is convenient for the creation of schematics, for example, for a lab report, presentation, etc.

This book contains 67 sample simulations. A brief summary of book chapters is given bellow:

Chapter 1 introduces the TINA-TI and shows how it can be used to analyze electric circuits. Students who take/took electric circuits I/II course can use this chapter as a reference to learn how to solve an electric circuit problem with the aid of computer. This chapter has 32 sample simulations.

Chapter 2 focus on the simulation of electronic circuits (i.e., circuits which contain diode, transistor, IC's, etc.) with TINA-TI. Students who take/took electronic I/II course can use this chapter as a reference to learn how to analyze an electronic circuit with the aid of computer. This chapter has 35 sample simulations.

I hope that this book will be useful to the readers, and I welcome comments on the book.

Farzin Asadi
(farzinasadi@maltepe.edu.tr)

List of Tables

Table 1.1 Available prefixes in TINA-TI.   19

1

Simulation of Electric Circuits with TINA-TI®

1.1 Introduction

In this chapter, you will learn how to analyse electric circuits in TINA-TI software. The theory behind the studied circuits can be found in any standard circuit theory textbook [1–4]. It is a good idea to do some hand calculations for the circuits that are given and compare them with TINA-TI results.

1.2 Installation of TINA-TI

TINA-TI can be downloaded from www.ti.com/tool/TINA-TI. The download page can be obtained by searching for 'tina ti download' in Google as well (Figure 1.1).

Q tina ti download   ✕ 🎤

Figure 1.1

Go to the download page and click the Downloads button (Figure 1.2).

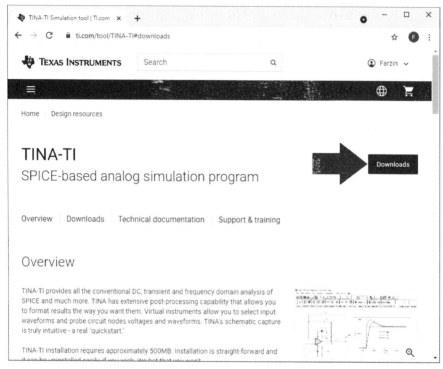

Figure 1.2

After clicking the Downloads button, the window shown in Figure 1.3 appears and asks you to log in. You need to have Texas Instruments account in order to be able to download the TINA-TI. If you don't have an account on the Texas Instruments website, you can make one by clicking the Register now link (Figure 1.3) and fill the form.

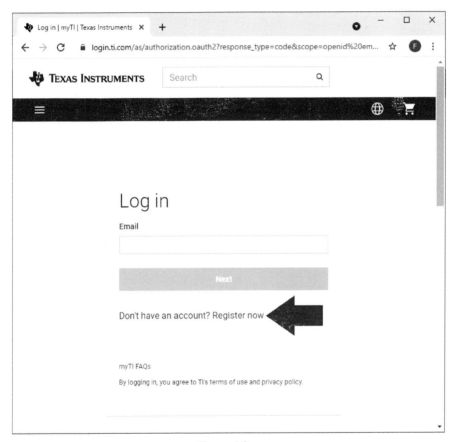

Figure 1.3

Download page of TINA-TI is shown in Figure 1.4. TINA-TI supports four different languages: English, Japanese, Russian and Chinese. Download and install your preferred one. This book used the English version.

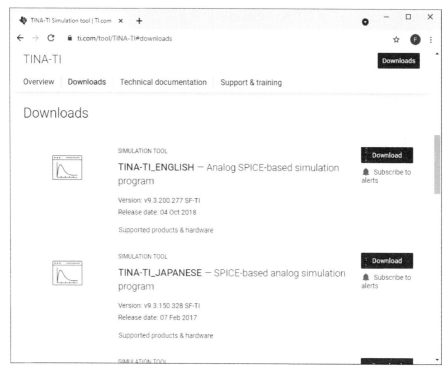

Figure 1.4

If you scroll down the download page you will see the Technical Documentation section (Figure 1.5). This section contains useful materials. Don't lose it.

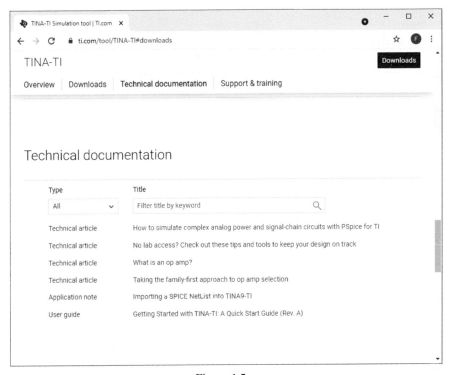

Figure 1.5

Texas Instrument produced some video tutorials for TINA-TI. These videos can be accessed in https://training.ti.com/tina-ti-tutorial page (Figure 1.6).

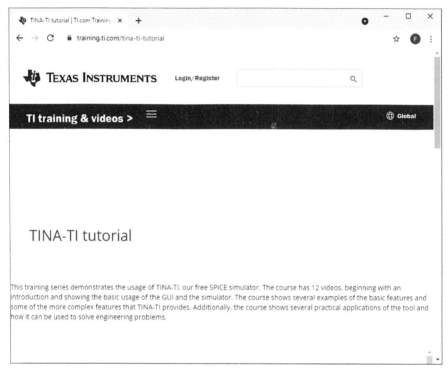

Figure 1.6

TINA-TI has some ready to use sample simulations. Theses samples can be accessed by clicking the File> Open Examples (Figure 1.7).

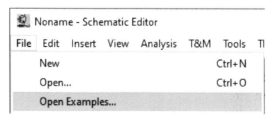

Figure 1.7

1.3 Version of Software

If you click the Help> About (Figure 1.8), version of software is shown (Figure 1.9).

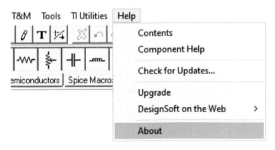

Figure 1.8

Figure 1.9

1.4 Example 1: Simple Resistive Voltage Divider

In this example we want to simulate the resistive voltage divider shown in Figure 1.1 resistive circuit From basic circuit theory we know that $V_{R1} = \frac{R_1}{R_1+R_L} \times V_{in} = \frac{1k}{1k+2.2k} \times 10 = 3.03\ V$ and $V_{RL} = \frac{R_L}{R_1+R_L} \times V_{in} = \frac{2.2k}{1k+2.2k} \times 10 = 6.97\ V$. The current drawn from DC source is $I = \frac{V_{in}}{R_1+R_L} = \frac{10}{1k+2.2k} = 3.125\ mA$.

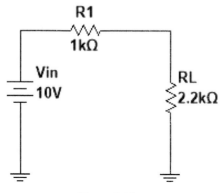

Figure 1.10

Let's simulate this circuit. Open the software (Figure 1.11). TINA-TI has some grid dots on the screen by default. The grid dots can be removed by clicking the Grid on/off the icon (Figure 1.12).

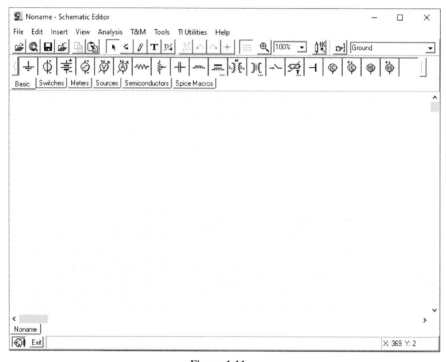

Figure 1.11

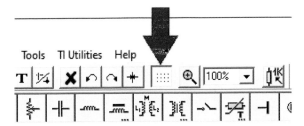

Figure 1.12

Click the Basic tab and add the required components to the schematic (Figure 1.13). Note that Battery and Voltage Source blocks (Figure 1.14) are the same. So, you can use a Voltage Source block as well. You can use the block shown in Figure 1.15 if your schematic needs an independent DC current source.

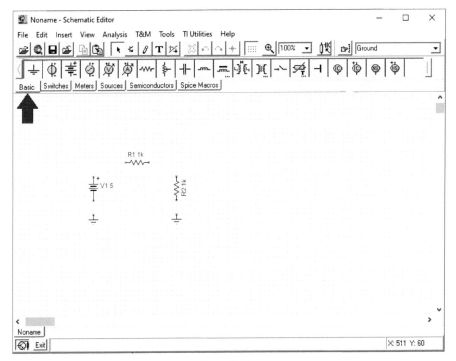

Figure 1.13

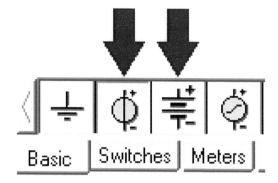

Figure 1.14

Figure 1.15

You can rotate a component clockwise by pressing the Ctrl+R or + key of your keyboard. You can rotate a component counterclockwise by pressing the – key of your keyboard. You can rotate a component with the aid of icons shown in Figure 1.16 as well.

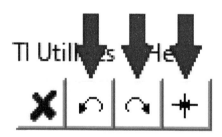

Figure 1.16

If you select the All from the Zoom, drop down list (Figure 1.17), the unused sections of schematic are removed and you obtain a better view of schematic (Figure 1.18).

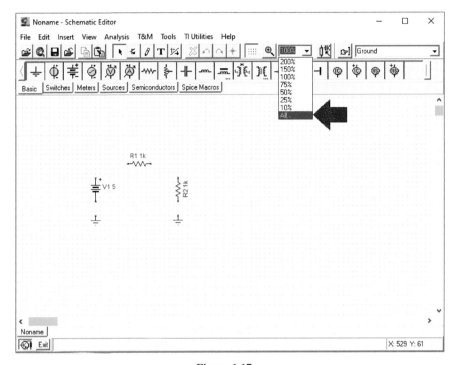

Figure 1.17

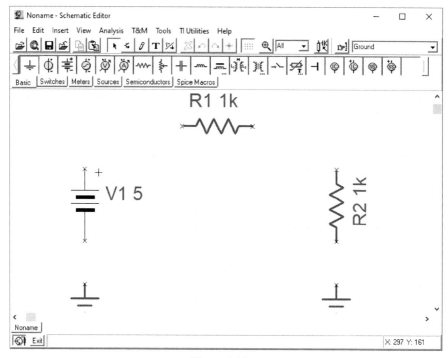

Figure 1.18

Click the Wire icon (Figure 1.19) and connect the terminals together (Figure 1.20).

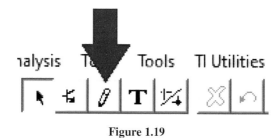

Figure 1.19

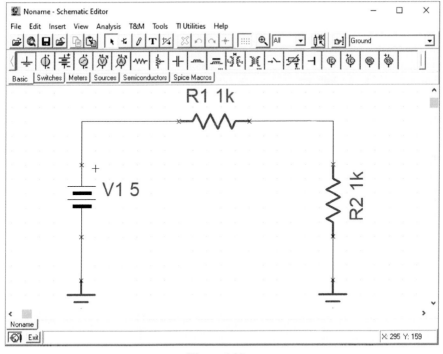

Figure 1.20

You can remove a wire (or a component) by right clicking on it and clicking the Delete (Figure 1.21).

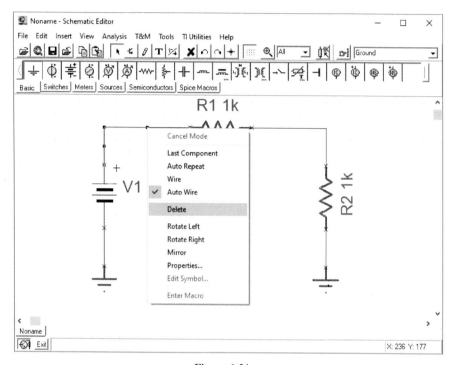

Figure 1.21

Another way to remove a wire (or a component) is left clicking on it (Figure 1.22) and pressing the Delete key of keyboard.

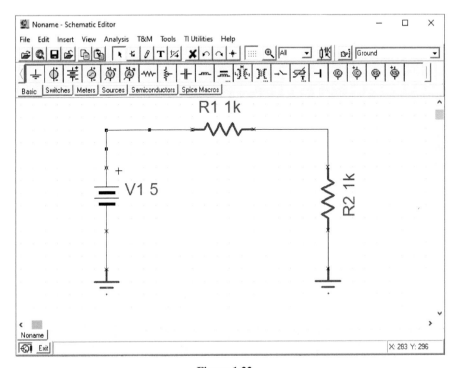

Figure 1.22

User can assign desired names to the circuit nodes. For instance, in Figure 1.23 the user assigned name 'node 1' to one of nodes.

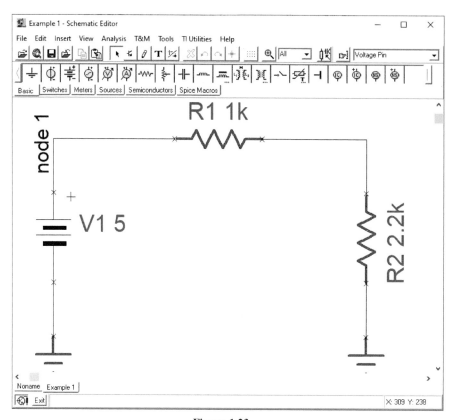

Figure 1.23

The user can assign a name to a node by double-clicking one of the wires which is connected to it. After double-clicking the wire, the window shown in Figure 1.24 appears. Enter the desired name in the ID box. Uncheck the Show ID box if you don't want to show the label name on the schematic.

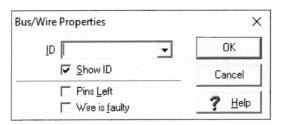

Figure 1.24

It is time to enter the values of components. Double click the Battery block and enter Vin and 10 to the Label and Voltage (V) boxes, respectively (Figure 1.25). Then click the OK button.

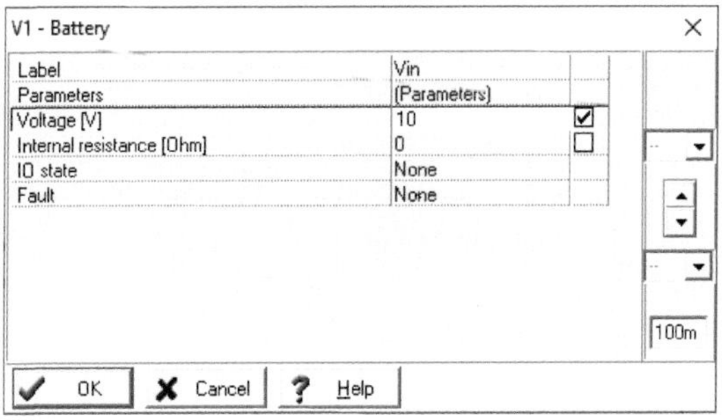

Figure 1.25

After clicking the OK button, the schematic shows the new name and value (Figure 1.26).

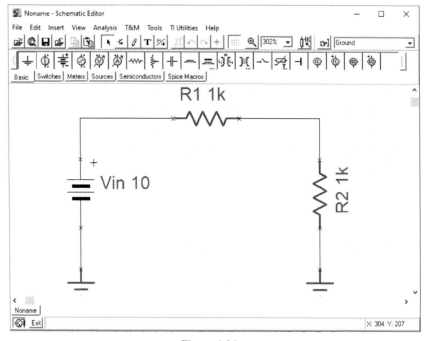

Figure 1.26

Double click the resistor R2 and enter RL and 2.2k to the Label and Resistance (Ohm) boxes, respectively (Figure 1.27). Then click the OK button. 2.2k means 2.2 kΩ. Note that you can use the prefixes shown in Table 1.1 when you want to enter the value of components.

RL - Resistor			✕
Label	RL		
Parameters	(Parameters)		
Resistance [Ohm]	2.2k	...	☑
Power [W]	1		☐
Temperature	Relative		
Temperature [C]	0		☐
Linear temp. coef. [1/C]	0		☐
Quadratic temp. coef. [1/C²]	0		☐
Exponential temp. coef. [%/C]	0		☐
Maximum voltage (V)	100		☐
Fault	None		

✓ OK   ✕ Cancel   ? Help

Figure 1.27

Table 1.1   Available prefixes in TINA-TI.

Prefix	Meaning
G	Giga
Meg	Mega
k	Kilo
m	Milli
u	Micro
n	Nano
p	Pico
f	Femto

If you click the Help button in Figure 1.27, the window shown in Figure 1.28 appears on the screen and shows the description of resistor parameters. The help of software is a good reference to see the details of blocks.

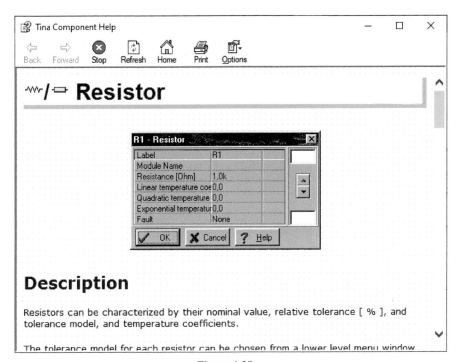

Figure 1.28

You can use the text icon (Figure 1.29) to add text to the schematic (Figure 1.29). You can remove the added text by clicking on it and pressing the Delete key on your keyboard.

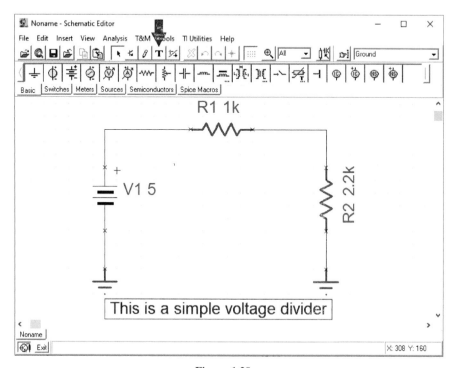

Figure 1.29

Now we need to determine the variable(s) that we look for. We want to find the voltage of resistor R2. Click the Meters tab and add a Voltage Pin block to the schematic (Figure 1.30).

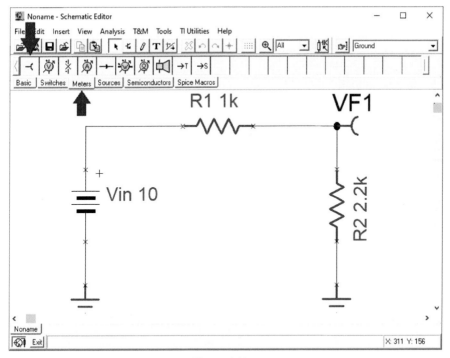

Figure 1.30

Double click the Voltage Pin block and enter-output to the Label box (Figure 1.31). Then click the OK button. The schematic changes to what are shown in Figure 1.32.

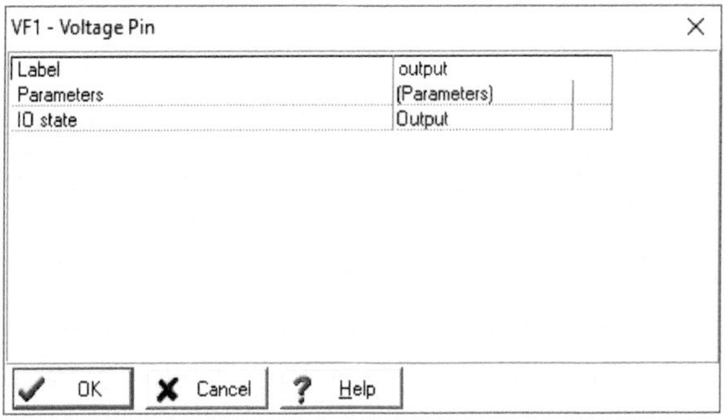

Figure 1.31

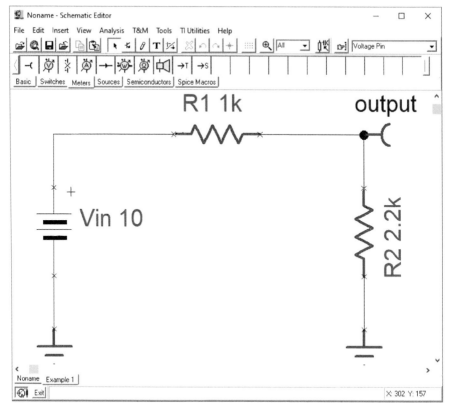

Figure 1.32

You can use the File> Save to save the schematic (Figure 1.33).

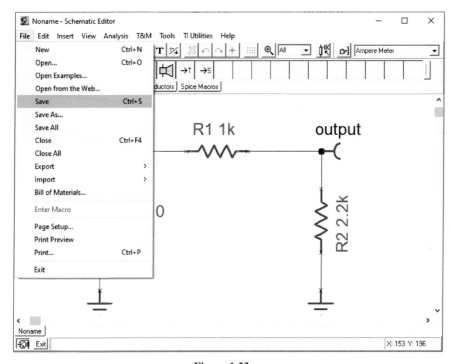

Figure 1.33

The schematic drawing in TINA-TI can be exported as a graphical file. This is very useful when you want to write a report and you need to show the circuit. In order to export the drawn schematic as a graphical file, click the File> Export> Picture File (*.EMF; *.BMP; *.PDF) (Figure 1.34) and determine the file name and save path.

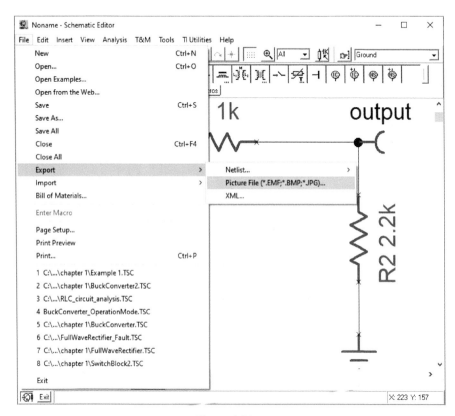

Figure 1.34

Click the Analysis> ERC (Figure 1.35). This runs the Electrical Rules Check (ERC) and display the results in the Electrical Rules Check window (Figure 1.36). ERC will examine the circuit for questionable connections between components based on the ERC Matrix. According to Figure 1.36, our schematic has 0 errors and 0 warnings. So, we can continue.

You can do the simulation even in presence of warnings. However, it is recommended to check your schematic and see why you received warnings.

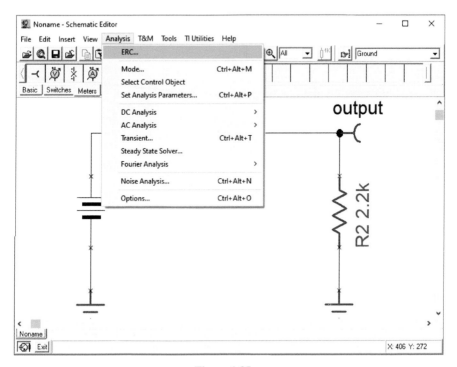

Figure 1.35

Figure 1.36

The ERC Matrix can be seen by clicking the Analysis> Options (Figure 1.37). After clicking the Analysis> Options, the Analysis Options window appears on the screen. Click the ERC tab to see the ERC Matrix (Figure 1.38). In this section, 'E' shows error and 'W' shows a warning. Click the Help button in Figure 1.38 to obtain more information about this tab.

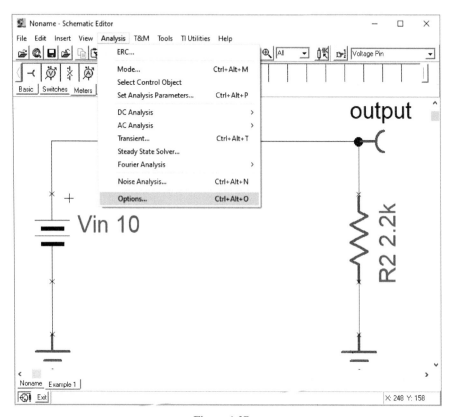

Figure 1.37

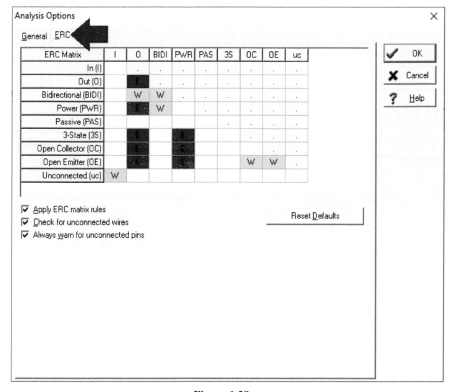

Figure 1.38

Let's run the simulation. Click the Analysis> DC Analysis> Calculate nodal voltages (Figure 1.39). Result is shown in Figure 1.40 According to Figure 1.40, voltage of node output is 6.88 V.

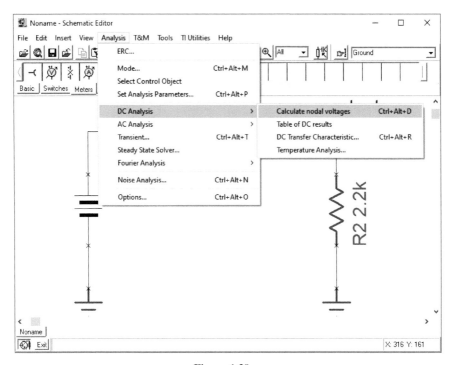

Figure 1.39

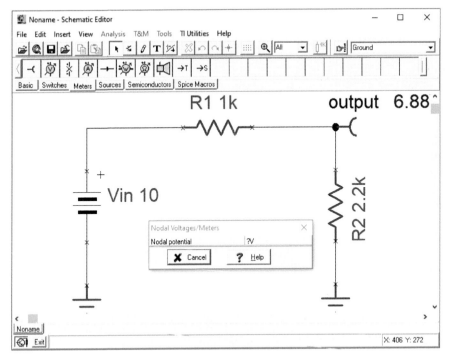

Figure 1.40

If you click on the circuit components, their voltage and current will be shown (Figure 1.41). Note that after clicking on a component an arrow appears on it (Figure 1.42). This arrow shows the direction of the current measurement. So, the value that is shown by TINA-TI is the current that goes from A to B. The voltage that is shown by TINA-TI is the voltage difference between points A and B, i.e., $V_{AB} = V_A - V_B$. You can finish the simulation by clicking the Cancel button in Figure 1.40 or Figure 1.41.

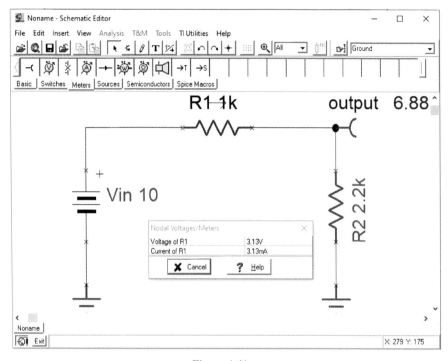

Figure 1.41

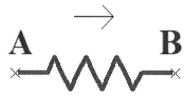

Figure 1.42

TINA-TI can show the analysis results in tabular format as well. Click the Analysis> DC Analysis> Table of DC results in order to see the analysis result in tabular format. After clicking, the schematic changes to what is shown in Figure 1.44 and the results shown in Figure 1.45 appear on the screen. Note that TINA-TI automatically assigns a name to the nodes that are not named by the user. According to Figure 1.44, TINA-TI assigned name 2 to the positive terminal of voltage source Vin (=left terminal of resistor R1) and 0 to the ground.

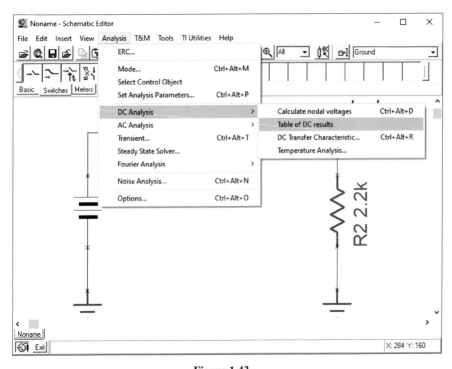

Figure 1.43

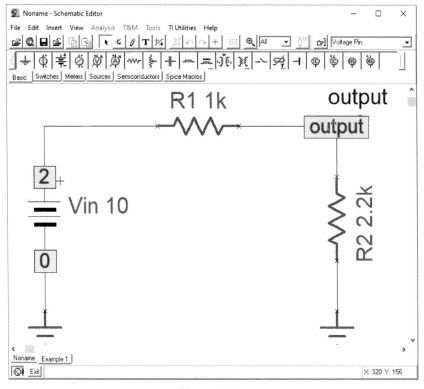

Figure 1.44

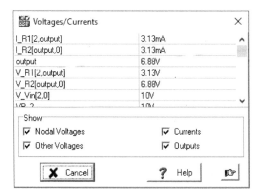

Figure 1.45

If you click any node or any component in the schematic, related data in the table is highlighted. For instance, if you click the R1, the rows shown in Figure 1.46 are highlighted. I_R1(2, output) =3.13 mA means that current through resistor R1 (current that goes from node '2' to node 'output') is 3.13 mA. V_R1(2, output) =3.13 V means voltage across resistor R1 ($V_{node\ "2"}$ − $V_{node\ "output"}$) is 3.13 V.

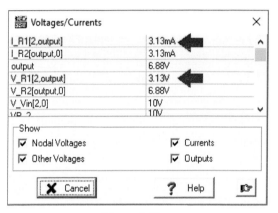

Figure 1.46

You can save the simulation result as a .txt file by clicking the hand icon (Figure 1.47).

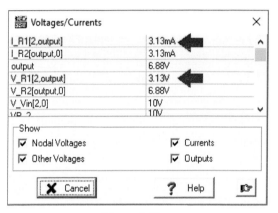

Figure 1.47

You can finish the simulation by clicking the Cancel button in Voltage/Currents window (Figure 1.47).

1.5 Example 2: Volt Meter and Ampere Meter Blocks

Voltmeter and Ampere meter blocks (Figure 1.48) are introduced in this example. The internal resistance of Voltmeter and Ampere meter blocks are $\infty\Omega$ and 0Ω, respectively.

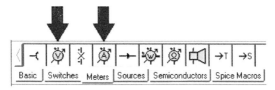

Figure 1.48

Draw the schematic shown in Figure 1.49.

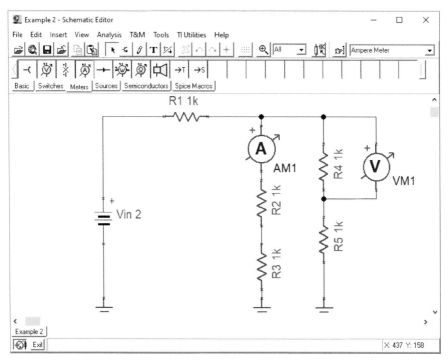

Figure 1.49

Double click the Ampere meter AM1 and change the Label box to I1 (Figure 1.50). Then click the OK button.

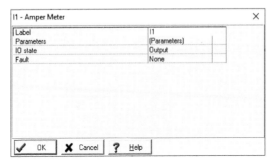

Figure 1.50

Double click the volt meter VM1 and change the Label box to VR4 (Figure 1.51). Then click the OK button.

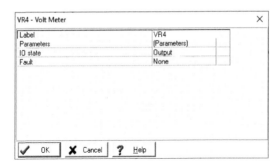

Figure 1.51

The schematic changes to what shown in Figure 1.52.

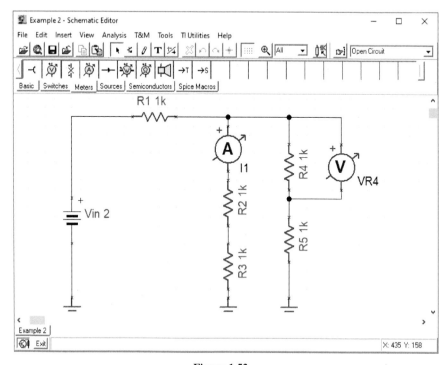

Figure 1.52

Click the Analysis> DC Analysis> Calculate nodal voltages (Figure 1.53). After clicking, the result shown in Figure 1.54 appears on the screen. According to Figure 1.54, the voltmeter reading is 500 mV and the Ampere meter reading is 500 μA. Use hand analysis to ensure that the obtained result is correct.

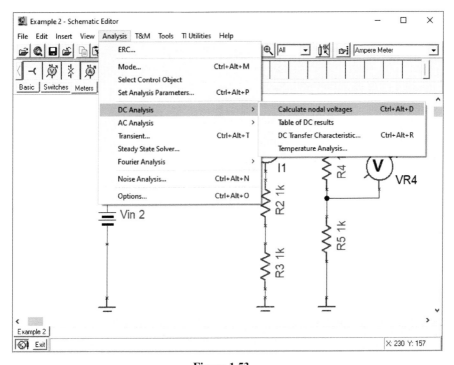

Figure 1.53

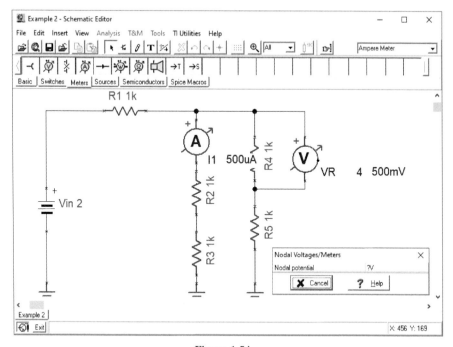

Figure 1.54

Stop the simulation by clicking the Cancel button in Figure 1.54. Add a capacitor with a capacitance of 1 μF in series with the input voltage source and re-run the simulation (Figure 1.55). This time the Voltmeter and Ampere meter blocks read 0 V and 0 A, respectively. Remember that a capacitor acts like an open circuit in steady-state DC conditions. Therefore, there is an open circuit in the input of the circuit and no current flows in a steady state. That is why the Voltmeter and Ampere meter blocks read 0 V and 0 A.

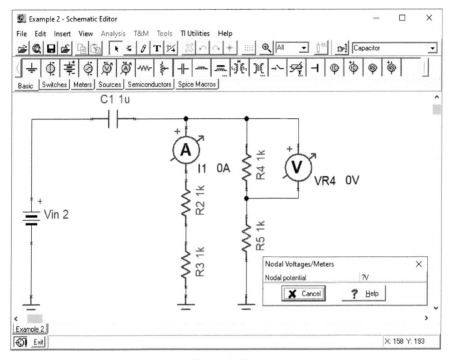

Figure 1.55

1.6 Example 3: Open Circuit and Current Arrow Blocks

In the previous example, we used the Voltmeter and Ampere meter blocks to measure the circuit voltage and current. The open circuit and current arrow blocks (Figure 1.56) can be used to measure the voltage and current as well. In this example, we will simulate the previous example with the aid of an open circuit and current arrow blocks.

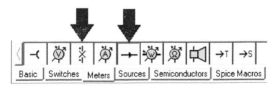

Figure 1.56

Change the schematic of Example 2 to what shown in Figure 1.57.

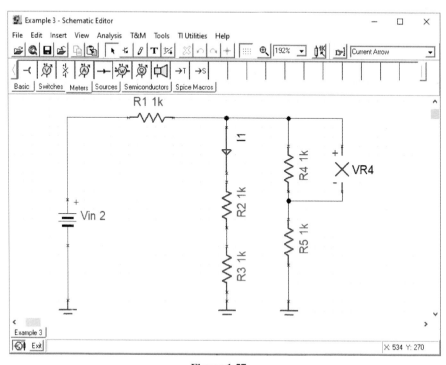

Figure 1.57

Click the Analysis> DC Analysis> Calculate nodal voltages to run the simulation. Simulation result is shown in Figure 1.58. Obtained result is the same as previous example result (Figure 1.54).

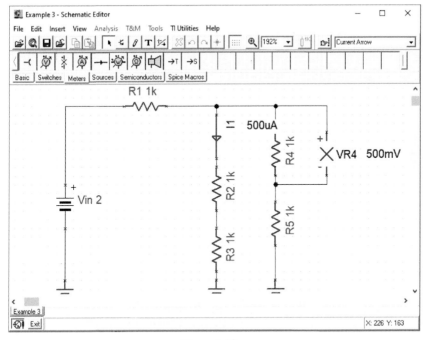

Figure 1.58

1.7 Example 4: RLC Circuit with Non-zero Initial Condition

In this example, we want to simulate the RLC circuit shown in Figure 1.59. An Initial current of the inductor and initial voltage of the capacitor is 2 A and 25 V, respectively. We want to study this circuit in the [0, 3s] interval.

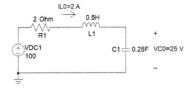

Figure 1.59

Let's simulate the circuit in TINA-TI. Draw the schematic shown in Figure 1.60.

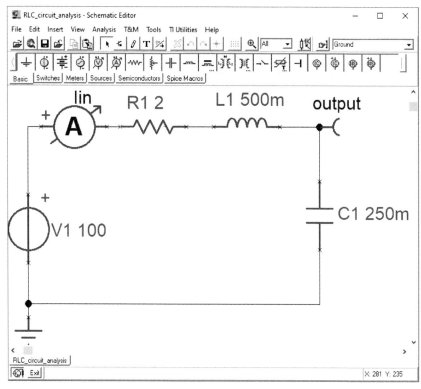

Figure 1.60

Double click the inductor and enter the given initial current to the Initial DC current (A) box (Figure 1.61).

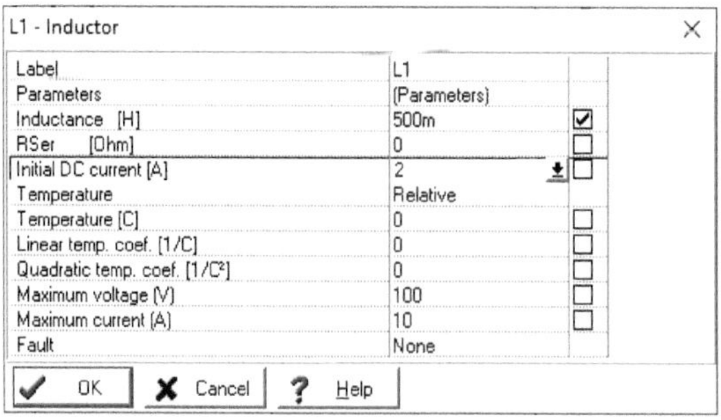

Figure 1.61

Double click the capacitor and enter the given initial voltage to the Initial DC voltage (V) box (Figure 1.62).

C1 - Capacitor		×
Label	C1	
Parameters	(Parameters)	
Capacitance [F]	250m	☑
RPar [Ohm]	Infinite	☐
Initial DC voltage [V]	25	±☐
Temperature	Relative	
Temperature [C]	0	☐
Linear temp. coef. [1/C]	0	☐
Quadratic temp. coef. [1/C²]	0	☐
Maximum voltage [V]	100	☐
Maximum ripple current [A]	1	☐
Fault	None	

✓ OK	✗ Cancel	? Help

Figure 1.62

Note that when you click the inductor or capacitor icons from the toolbar and you want to add them to the schematic, the mouse pointer is attached to one of the terminals. Let's call the terminal which is attached to the mouse pointer '+' and the other terminal '-' (Figure 1.63).

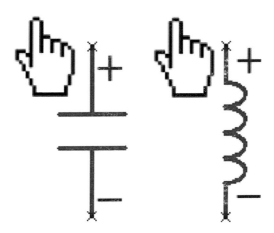

Figure 1.63

When you enter 2 into the Initial DC current (A) box of inductor L1, it means that 2 A enters the + terminal of inductor L1. When you enter 25 into the Initial DC voltage (V) box of capacitor C1, it means that $V_+ - V_- = 25\ V$ (V_+ and V_- show the + terminal voltage and negative terminal voltage, respectively). The '+' terminals of inductor and capacitor are shown in Figure 1.64.

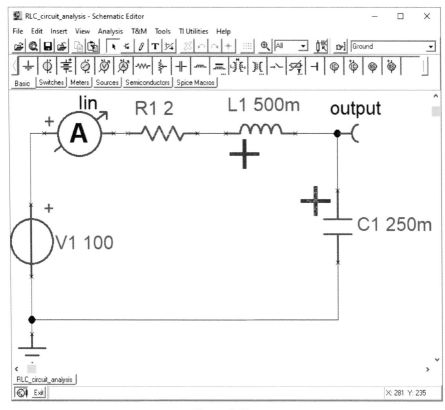

Figure 1.64

Click the Analysis> Set Analysis Parameters (Figure 1.65). This opens
the Analysis Parameters window (Figure 1.66).

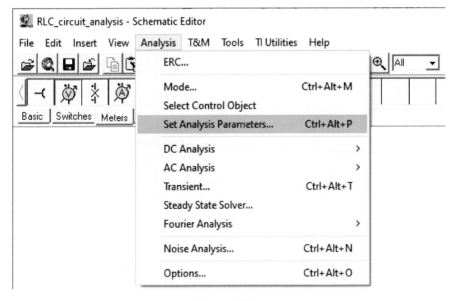

Figure 1.65

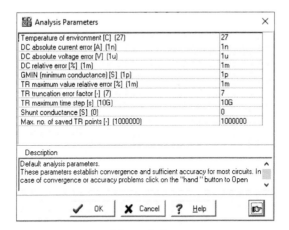

Figure 1.66

Click the hand icon. This opens a list for you. Click the View All (Figure 1.67).

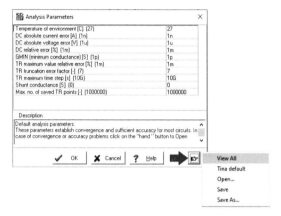

Figure 1.67

Now you can see all the parameters. Ensure that 'Operating point with initial conditions (No)' row is filled with 'Yes' (Figure 1.68).

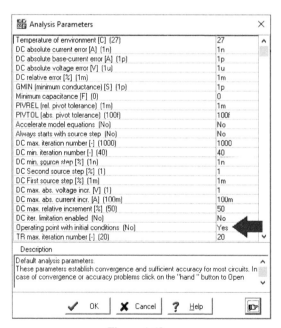

Figure 1.68

Click the Analysis> Transient (Figure 1.69). This opens the Transient Analysis window. Enter the desired time interval into the Start display and End display boxes. In this example, we want to study the circuit behavior on the [0, 3s] time interval. So, Start display=0 and End display=3 (Figure 1.70). Ensure that calculated operating point is selected.

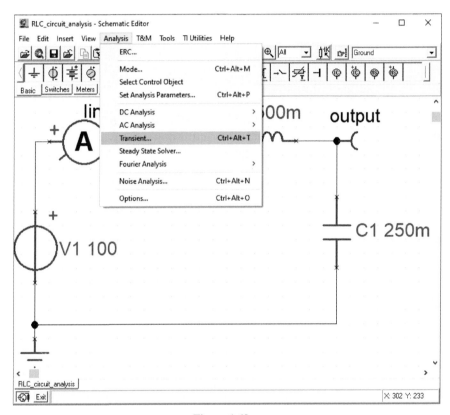

Figure 1.69

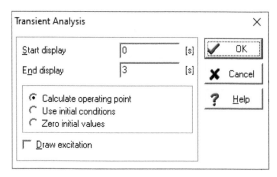

Figure 1.70

If you click the Help button in Figure 1.70, the help page of the block appears on the screen (Figure 1.71). Scroll down the page to see the description of the different types of available analysis (Figure 1.72).

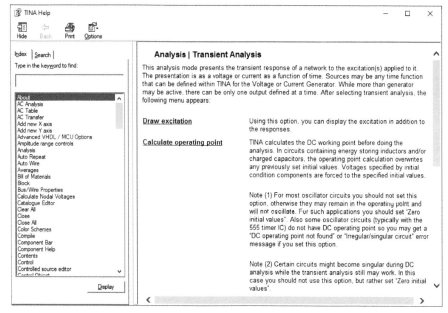

Figure 1.71

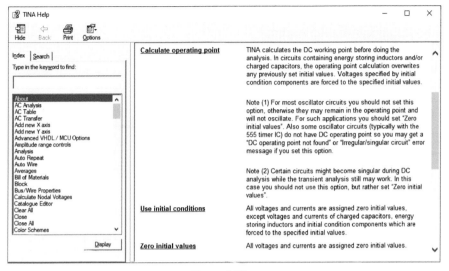

Figure 1.72

After clicking the OK button in Figure 1.70, the result shown in Figure 1.73 appears on the screen.

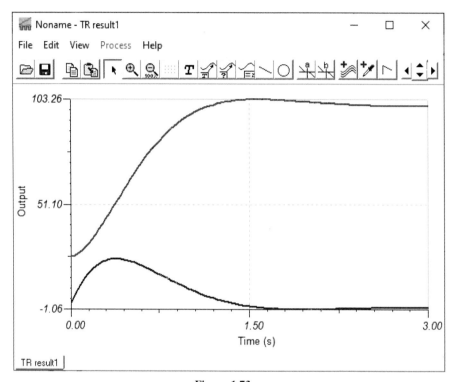

Figure 1.73

Click the Legend icon (Figure 1.74) to add a legend to the Figure. The legend helps you to understand which curve shows the circuit current and which curve shows the capacitor voltage. According to Figure 1.75, the red curve (lower curve) shows the circuit current and the green curve (upper curve) shows the capacitor voltage. You can remove any of the curves by clicking on it and pressing the Delete key on the keyboard.

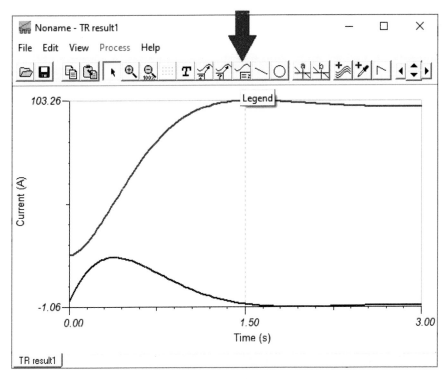

Figure 1.74

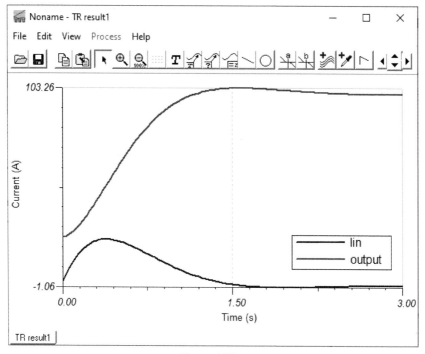

Figure 1.75

If you double click the curves, the window shown in Figure 1.76 appears and permits you to change the appearance (color, width and market type) of the curve.

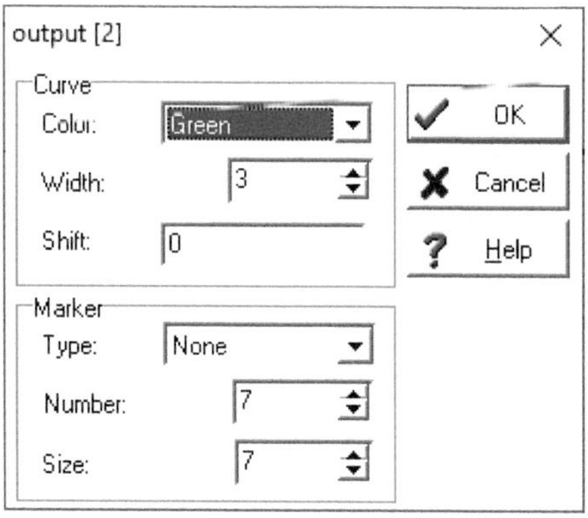

Figure 1.76

You can ask TINA-TI to draw the obtained curves on different coordinates. In order to do so, click the View> Separate curves (Figure 1.77). After clicking, the graph changes to what is shown in Figure 1.78.

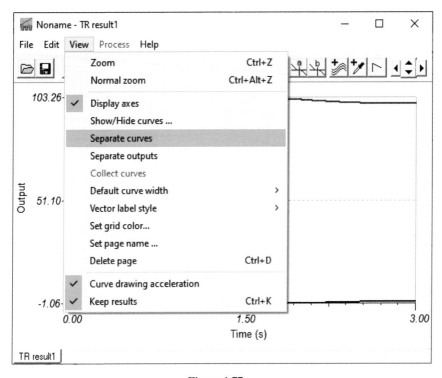

Figure 1.77

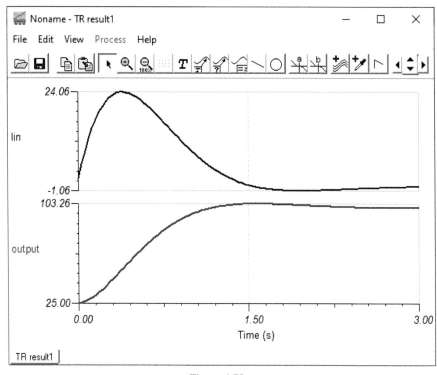

Figure 1.78

If you want to see the obtained curves on the same coordinate system, click the View> Collect Curves (Figure 1.79). After clicking, Figure 1.74 appears on the screen again.

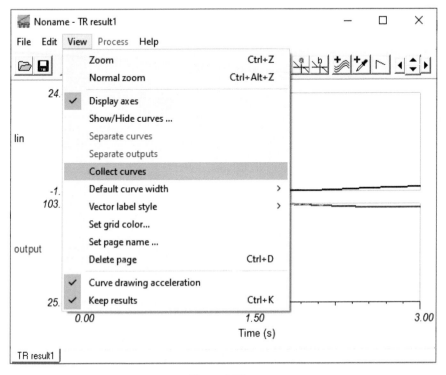

Figure 1.79

You can use the cursor icons (Figure 1.80) to read the curves. Let's ensure that the initial conditions of the curves are 2 A and 25 V. In order to do so, click one of the cursor icons. Then click on one of the curves. This assigns the cursor to the clicked curve. Drag the top of the cursor until it reaches the desired point. The coordinate of the point is shown in the opened box. According to Figure 1.80, both of the curves started from expected initial points.

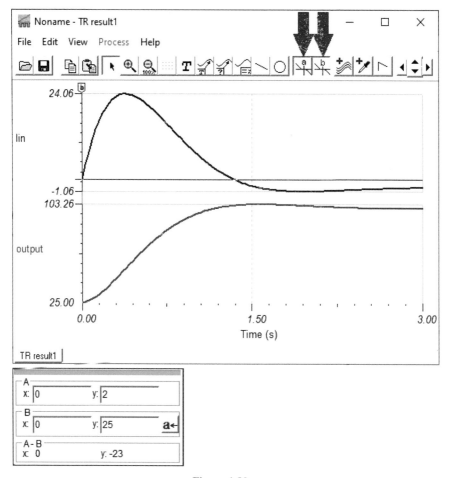

Figure 1.80

Let's use the cursor to measure the maximum of the obtained curves. According to Figure 1.81, the maximum circuit current is 24.05 A and it occurs at 384.44 ms. Maximum of the capacitor voltage is 103.26 V and it occurs at 1.55 s.

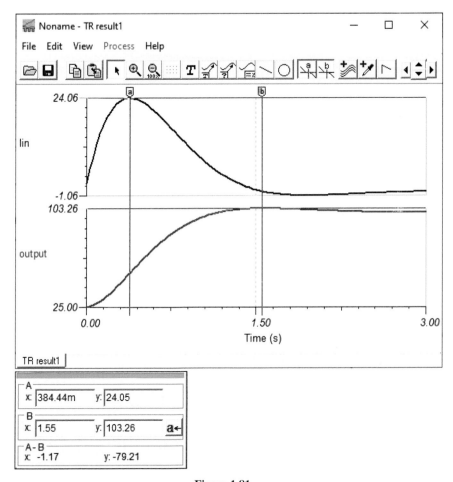

Figure 1.81

Let's check the obtained results. From basic circuit theory we have:
$$R.i\,(t) + L\frac{di(t)}{dt} + \frac{1}{C}\int_0^t i\,(\tau)\,d\tau + V_{0,C} = V_1 \Rightarrow R\frac{di(t)}{dt} + L\frac{d^2i(t)}{dt^2} + \frac{1}{C}i\,(t) =$$
$0 \Rightarrow \frac{d^2i(t)}{dt^2} + \frac{R}{L}\frac{di(t)}{dt} + \frac{1}{LC}i(t) = 0$. Initial conditions are $i_{0,L} = 2\ A$ and
$\frac{di(t)}{dt}\Big|_{t=0} = \frac{V_1 - Ri_{0,L} - V_{0,C}}{L} = \frac{100 - 2\times 2 - 25}{0.5} = 142\ \frac{A}{s}$. Following MATLAB

code solves the obtained differential equation and draws the graph of circuit current and the capacitor voltage over [0, 3s] time interval.

```
clc
clear all

R=2;
L=500e-3;
C=0.25;
V1=100;
a=R/L;
b=1/L/C;

syms i(t)
ode = diff(i,t,2)+a*diff(i,t)+b*i == 0;
Dy=diff(i,t);
cond = [i(0) == 2; Dy(0)==142];
iSol(t) = dsolve(ode,cond);
V0C=25;
syms x
VC=1/C*int(iSol,t,0,x)+V0C
VC=subs(VC,x,t)

figure(1)
ezplot(iSol,[0 3])
title('Circuit current (A)')
grid minor

figure(2)
ezplot(VC,[0 3])
title('Capacitor voltage(V)')
grid minor
```

The output of the code is shown in Figures 1.82 and 1.83. You can compare different points of these graphs with the TINA-TI result in order to ensure that the TINA-TI result is correct.

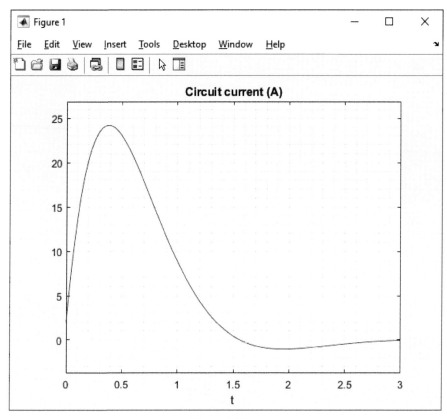

Figure 1.82

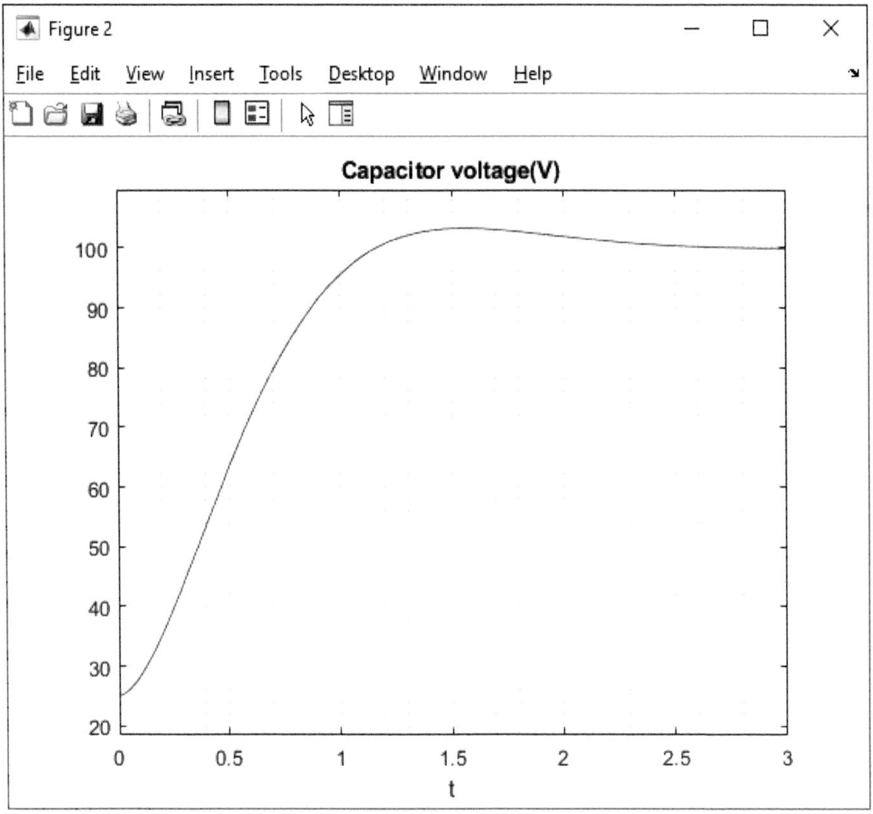

Figure 1.83

For instance, let's compare the maximum of obtained graphs with the TINA-TI results. According to Figure 1.84, the maximum circuit current is about 24.1972 A and it occurs at 381.062 ms. According to Figure 1.85, maximum of the capacitor voltage is 103.243 V and it occurs at 1.55889 s. These values are quite close to the TINA-TI result shown in Figure 1.81.

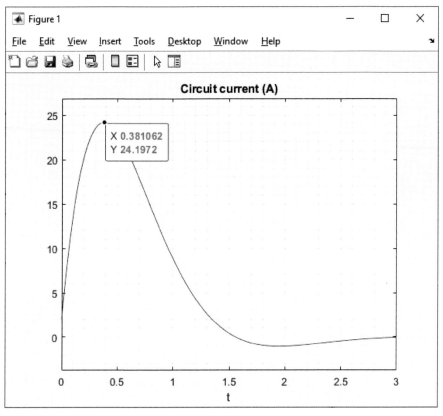

Figure 1.84

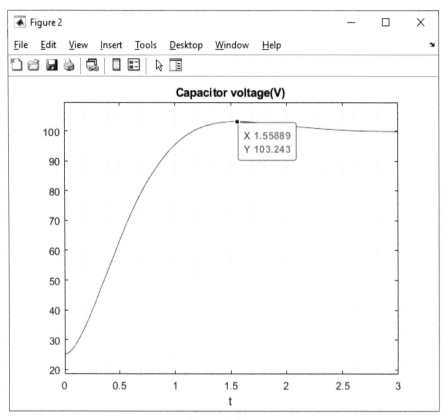

Figure 1.85

1.8 Example 5: Exporting the Obtained Waveforms as a Graphical File

The obtained waveforms can be exported as graphical files. This is very useful when you want to write a report and you need to show the circuit waveforms. In order to export the obtained waveforms as a graphical file, click the File> Export> Bitmap or File> Export> JPEG (Figure 1.86) and determine the file name and save path.

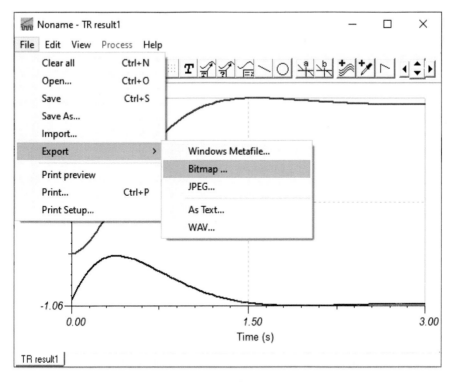

Figure 1.86

1.9 Example 6: Exporting the Obtained Waveforms as Text File

The obtained waveforms can be exported as text files. Exporting the wave-forms as text files permits further processing of the result. For instance, you can import the generated text file into MATLAB® or Excel® and use these software's to process the results.

Click the File> Export> As Text (Figure 1.87) in order to export the waveforms as a text files. After clicking, the Save diagram window (Figure 1.88) appears on the screen and asks for a save path and name. After entering the desired name, click the Save button.

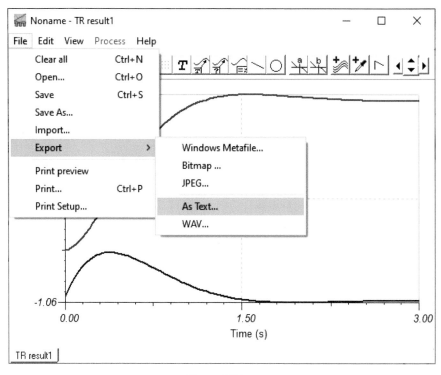

Figure 1.87

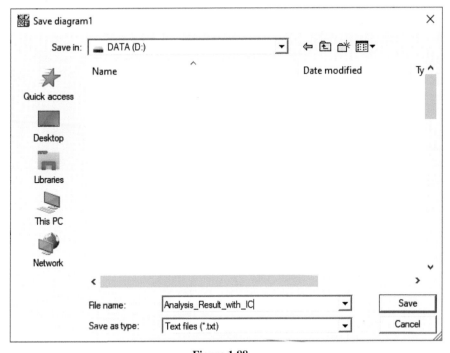

Figure 1.88

You can open the generated text file in Notepad (Figure 1.89).

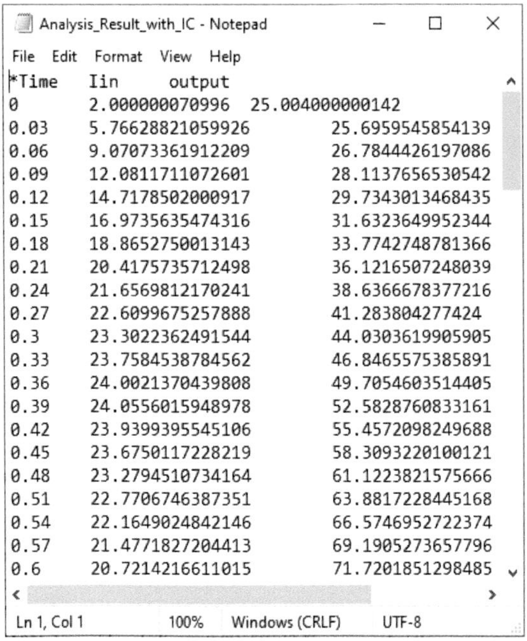

```
Analysis_Result_with_IC - Notepad                    —    □    ×

File   Edit   Format   View   Help
*Time     Iin        output
0         2.000000070996   25.004000000142
0.03      5.76628821059926        25.6959545854139
0.06      9.07073361912209        26.7844426197086
0.09      12.0811711072601        28.1137656530542
0.12      14.7178502000917        29.7343013468435
0.15      16.9735635474316        31.6323649952344
0.18      18.8652750013143        33.7742748781366
0.21      20.4175735712498        36.1216507248039
0.24      21.6569812170241        38.6366678377216
0.27      22.6099675257888        41.283804277424
0.3       23.3022362491544        44.0303619905905
0.33      23.7584538784562        46.8465575385891
0.36      24.0021370439808        49.7054603514405
0.39      24.0556015948978        52.5828760833161
0.42      23.9399395545106        55.4572098249688
0.45      23.6750117228219        58.3093220100121
0.48      23.2794510734164        61.1223821575666
0.51      22.7706746387351        63.8817228445168
0.54      22.1649024842146        66.5746952722374
0.57      21.4771827204413        69.1905273657796
0.6       20.7214216611015        71.7201851298485

<                                                        >
Ln 1, Col 1         100%    Windows (CRLF)     UTF-8
```

Figure 1.89

1.10 Example 7: RLC Circuit with Zero Initial Condition

In Example 4, we learned how to analyze a circuit with given initial conditions. You can ask TINA-TI to ignore the initial conditions and simulate the circuit with zero initial conditions. This example shows how to run the simulation with zero initial conditions.

Open the schematic of Example 4 and run it with the given initial conditions in Example 4. The result shown in Figure 1.90 appears on the screen.

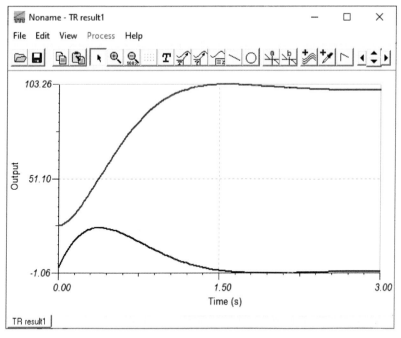

Figure 1.90

Close the window shown in Figure 1.90 and rerun the transient analysis. Select the Zero initial values in the Transient Analysis window (Figure 1.91) to run the simulation with zero initial conditions. The simulation result is shown in Figure 1.92. Note that the output window has two tabs and it holds the result of the previous analysis. You can move between tabs by clicking them.

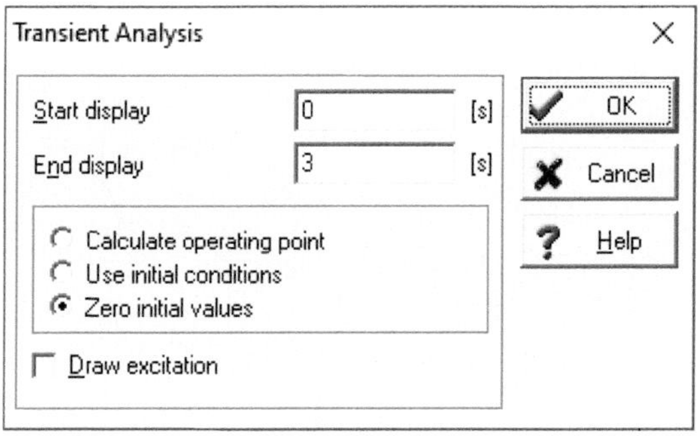

Figure 1.91

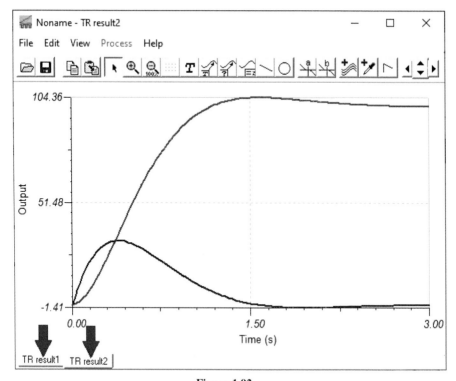

Figure 1.92

You can assign more meaningful names to the tabs. In order to do so, right-click and click the Set page name (Figure 1.93). This opens the Page Name window. Enter the desired name into the Enter page name box and click the OK button (Figure 1.94). After clicking the OK button, the entered name appears on the screen (Figure 1.95).

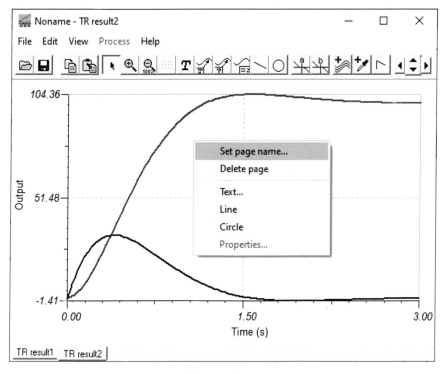

Figure 1.93

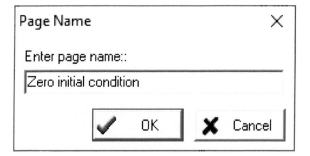

Figure 1.94

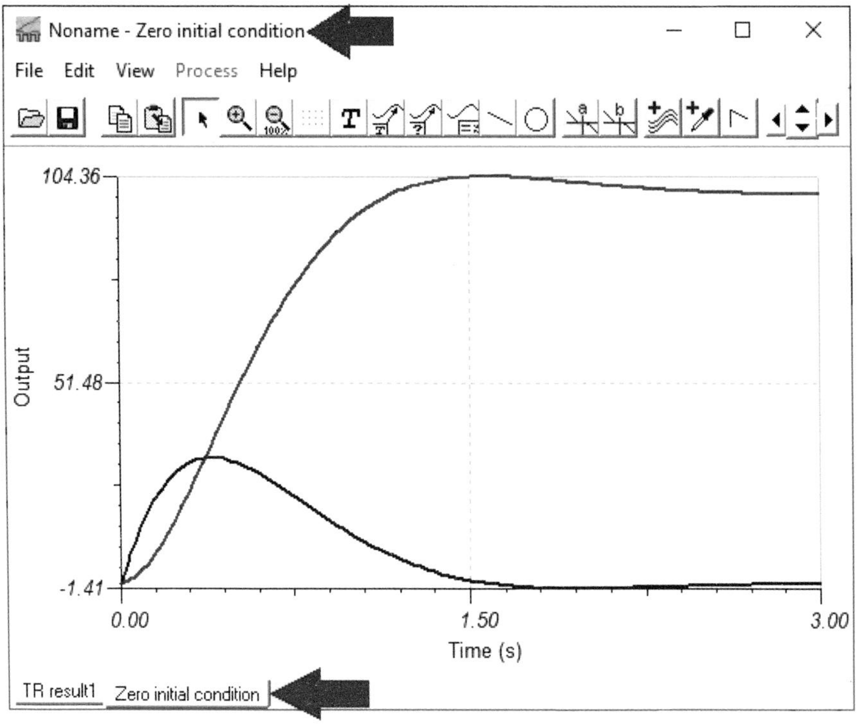

Figure 1.95

You can remove any of the available tabs by right-clicking on them and clicking the Delete page (Figure 1.96). You can remove all of the available tabs by clicking the File> Clear all (Figure 1.97).

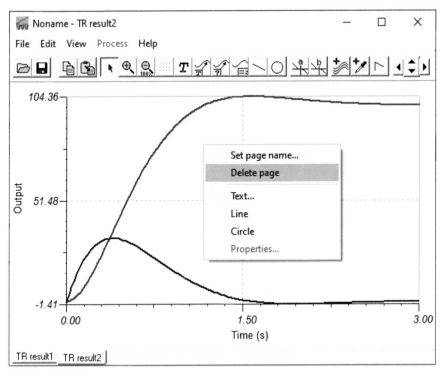

Figure 1.96

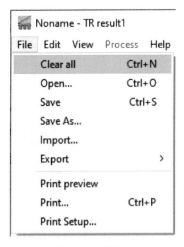

Figure 1.97

1.11 Example 8: Initial Condition Blocks

In Example 4 we entered the given initial voltage into the Initial DC voltage (V) box of the capacitor. In this example, we learn another method to enter the initial voltages to TINA-TI. In this example, we will use the Initial Condition 1 and Initial Condition 2 blocks (Figure 1.98). Using Initial Condition 1 you can specify a nodal voltage, while with Initial Condition 2 you can specify a voltage difference between two nodes.

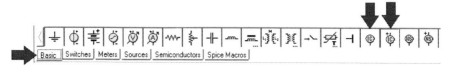

Figure 1.98

In this example, we want to simulate the RC circuit shown in Figure 1.99. The initial voltage of the capacitor is 2.5 V. We want to simulate the circuit for [0, 100 ms] time interval.

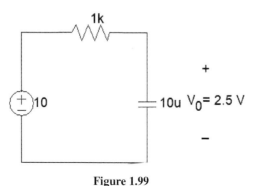

Figure 1.99

Draw the schematic shown in Figure 1.100 or Figure 1.101. Settings of capacitor C1, Initial Condition1 block IC1 and Initial Condition2 block IC2 are shown in Figure 1.102, Figure 1.103 and Figure 1.104, respectively.

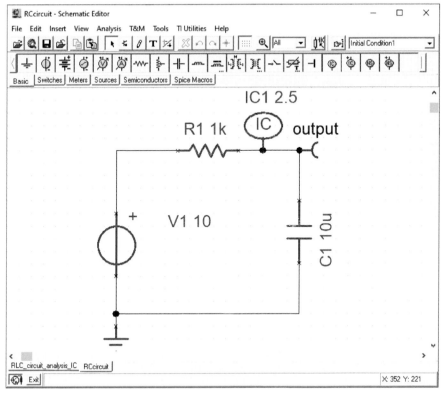

Figure 1.100

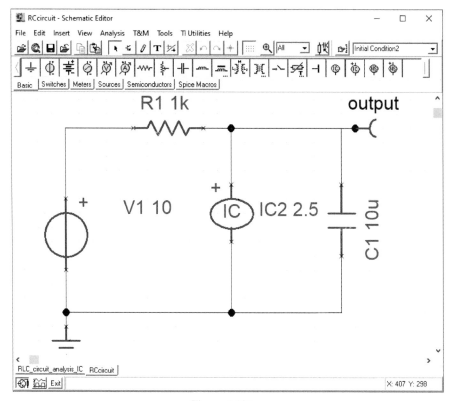

Figure 1.101

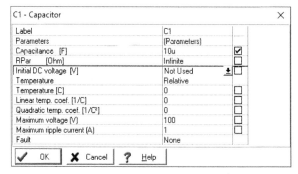

Figure 1.102

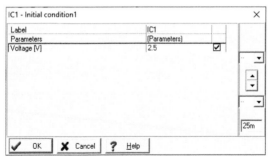

Figure 1.103

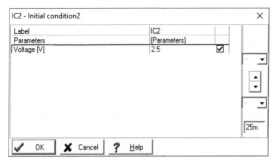

Figure 1.104

Run the transient analysis with the settings shown in Figure 1.105. The simulation result is shown in Figure 1.106. Note that the obtained waveform starts from 2.5 V and goes toward its final value of 10 V.

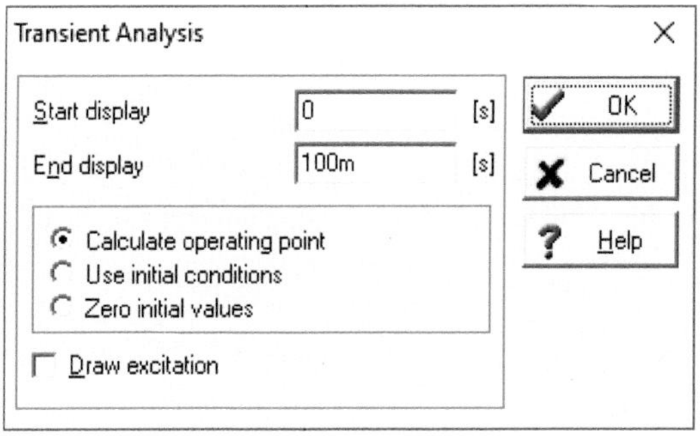

Figure 1.105

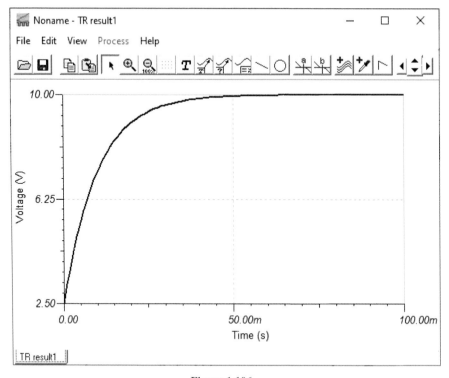

Figure 1.106

1.12 Example 9: Importing the TINA-TI Analysis Result into MATLAB®

In the previous example, we simulated a simple RC circuit with the given initial condition. In this example, we want to import the obtained result into MATLAB and find a mathematical equation for it.

Let's start. Click the File> Export> As Text (Figure 1.107) to generate the required text file. Exported file (tcurve.txt) is shown in Figure 1.108.

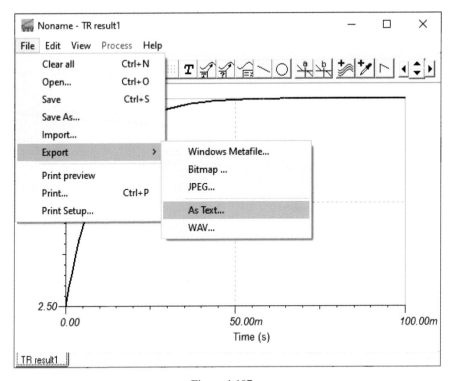

Figure 1.107

```
tcurve - Notepad                                    —    □    ×

File  Edit  Format  View  Help
*Time      output
0          2.50000037499998
0.001      3.18181852272726
0.002      3.80165320247932
0.003      4.38274821474689
0.004      4.91541864265882
0.005      5.39966448621512
0.006      5.83851228193802
0.007      6.2357452004803
0.008      6.59514641249474
0.009      6.92026264046833
0.01       7.21434504668081
0.011      7.48034923320465
0.012      7.72095371441806
0.013      7.93858300764613
0.014      8.13543072380201
0.015      8.3134812148631
0.016      8.47452941739074
0.017      8.62019889209372
0.018      8.75195817218254
0.019      8.87113556144889
0.02       8.97893252300407
0.021      9.07643579080231
0.022      9.16462832506411
0.023      9.24439922170442

Ln 1, Col 1         100%    Windows (CRLF)    UTF-8
```

Figure 1.108

Open the MATLAB and click the Import Data icon (Figure 1.109). After clicking, the Import Data window (Figure 1.110) appears on the screen. Use this window to open the text file that is generated by TINA-TI.

Figure 1.109

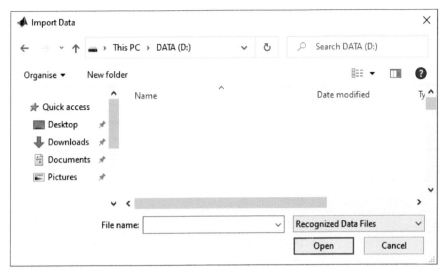

Figure 1.110

After opening the text file, the Import window (Figure 1.111) appears on the screen. Scroll down the screen. Note that the last row (row number 103) is Not a Number (NaN) (112).

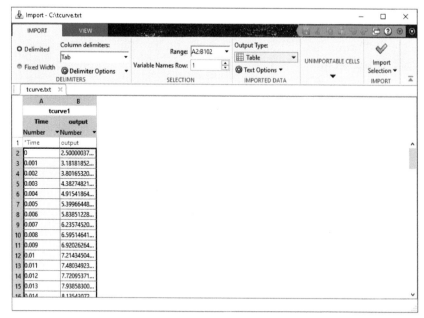

Figure 1.111

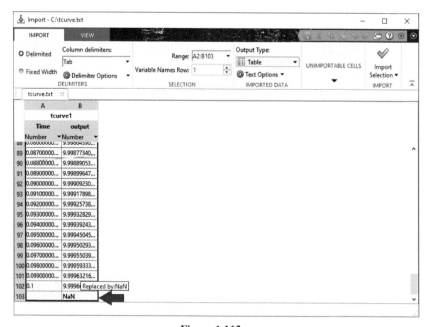

Figure 1.112

We need to get rid of 103 rows. In order to do so, click on the border and drag it to row 102 (Figure 1.113). Click the Import Selection button to enter the selected 102 rows into the MATLAB environment. After clicking a new variable is added to the MATLAB Workspace (Figure 1.114).

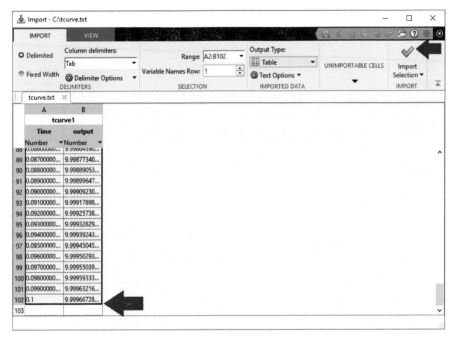

Figure 1.113

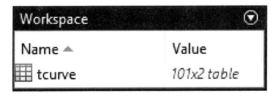

Figure 1.114

The commands shown in Figure 1.115 draw the graph of imported data. The output of this code is shown in Figure 1.116. This is exactly what we saw in Figure 1.106.

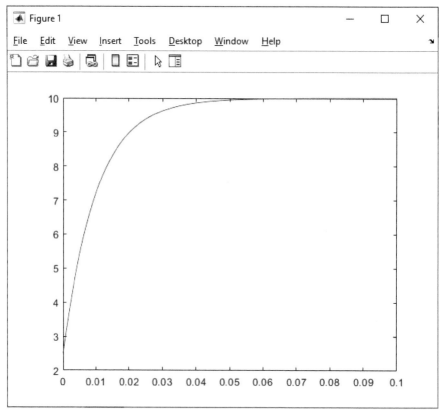

Figure 1.115

Figure 1.116

We solved the first part of our problem. It's time to use MATLAB to find a mathematical formula for the graph shown in Figure 1.116. The initial and final values of the graph shown in Figure 1.116 are 2.5 V and 10 V, respectively. From basic circuit theory we know that the form of an equation is $(I - F) e^{-\frac{t}{\tau}} + F$. I, F and τ show the initial value, final value and time constant of the circuit. So, the equation of the graph of Figure 1.116 is $(2.5 - 10)\,e^{-\frac{t}{\tau}} + 10$. The only unknown is τ. Let's use the Curve Fitting Toolbox to find the best value for τ. Run the toolbox by entering cftool into the MATLAB command prompt (Figure 1.117).

Figure 1.117

Do the settings similar to Figure 1.118. According to Figure 1.118, the best value of τ is 0.01009. Let's compare the obtained value with the correct one. In RC circuits time constant is defined as $\tau = R.C$. For $R = 1\ \text{k}\Omega$ and $C = 10\ \mu\text{F}$, the time constant is $\tau = R.C = 0.01$ s. This shows that MATLAB result is correct.

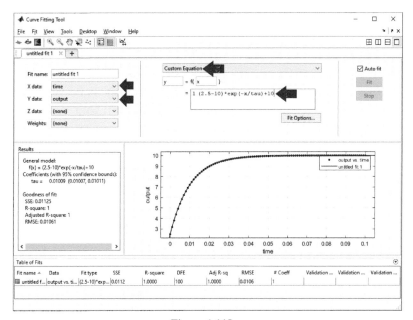

Figure 1.118

1.13 Example 10: Measurement of Phase Difference

In this example, we want to measure the phase difference between points B and A in Figure 1.119. The input voltage has a frequency of 60 Hz and a peak value of 1 V. From basic circuit theory, in steady state $V_B = \frac{j \times L \times \omega}{R + j \times L \times \omega} V_A =$ $\frac{j \times L \times 2\pi f}{R + j \times L \times 2\pi f} V_A = \frac{j \times 5m \times 377}{4 + j \times 5m \times 377} V_A = \frac{1.885j}{4 + 1.885j} V_A = 0.426 e^{j64.76°} V_A$. V_A and V_B show the phasor of the voltage of node A and B, respectively. So, the phase difference between points B and A is $64.76°$.

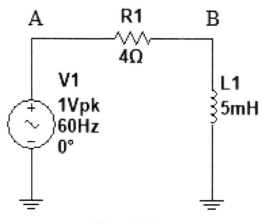

Figure 1.119

Let's solve this problem with TINA-TI. Draw the schematic shown in Figure 1.120. Required sinusoidal voltage is generated with a voltage generator block (Figure 1.121).

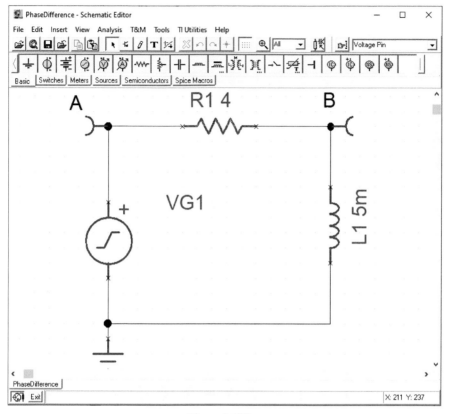

Figure 1.120

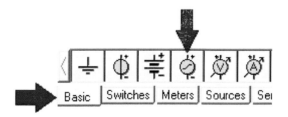

Figure 1.121

Double click the voltage generator block. The window shown in Figure 1.122 appears on the screen. Click the three dots in front of the Signal box. Then select the sinusoidal waveform and set the amplitude and frequency (Figure 1.123).

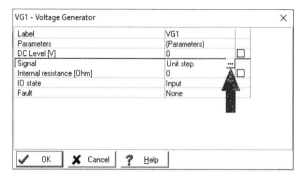

Figure 1.122

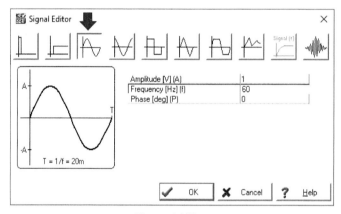

Figure 1.123

Run a transient analysis with the settings shown in Figure 1.124. Simulation result is shown in Figure 1.125.

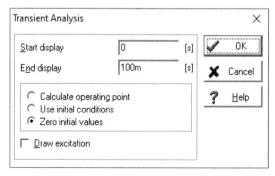

Figure 1.124

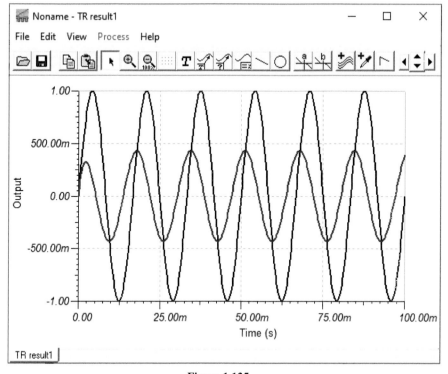

Figure 1.125

Use cursors to measure the time difference between the starting point of two waveforms (Figure 1.126). According to Figure 1.127, the time difference between the starting point of two waveforms is 2.99 ms.

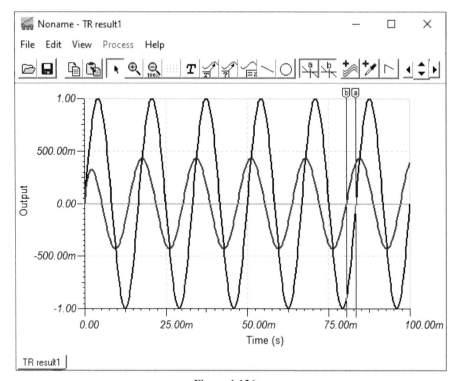

Figure 1.126

Figure 1.127

Calculations are shown in Figure 1.128 convert the measured time differ-ence into degrees. According to Figure 1.128, the phase difference between the two waveforms is 64.5840°.

```
Command Window                        ⊙
   >> T=1/60;
   >> Delta=2.99e-3;
   >> DeltaPhi=Delta/T*360

   DeltaPhi =

        64.5840

fx >>
```

Figure 1.128

1.14 Example 11: Power Meter Block

In this example, we want to measure the power of the circuit shown in Figure 1.129. The input voltage of this circuit is $120\sqrt{2}\sin(2 \times \pi \times 60 \times t + 45°) \cong 169.7\sin(2 \times \pi \times 60 \times t + 45°)$. Measurement of power is done with the aid of a Power Meter block (Figure 1.130). The Power Meter consists of a voltmeter and an ammeter. The voltmeter is represented by the thin vertical line, while the ammeter is indicated by the heavy horizontal line. The positive end of the reference direction (i.e., the starting end of the arrow) is referred to by' + '. Note that this block gives correct results in linear circuits. It does not generate correct results in nonlinear circuits (i.e. when voltage/current waveform contains harmonics). The Power Meter block can be used in DC circuits as well.

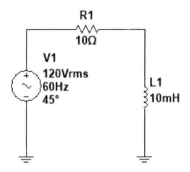

Figure 1.129

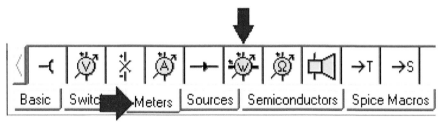

Figure 1.130

Let's start. Draw the schematic shown in Figure 1.131. The current of this circuit doesn't contain any harmonics. Therefore, the Power Meter block generates correct results in this circuit.

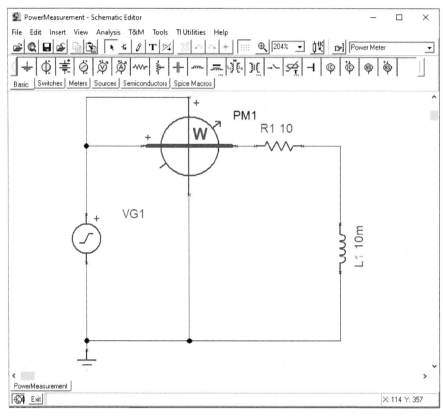

Figure 1.131

Settings of input voltage source is shown in Figure 1.132.

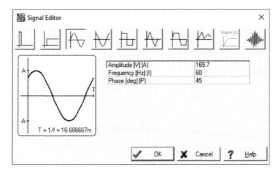

Figure 1.132

Click the Analysis> AC Analysis> Calculate nodal voltages (Figure 1.133). Then click on the Power Meter block PM1. After clicking the Power Meter block PM1, the result shown in Figure 1.134 appears on the screen.

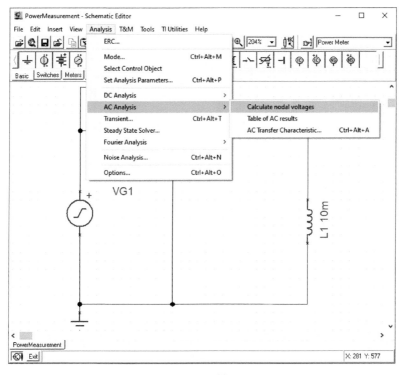

Figure 1.133

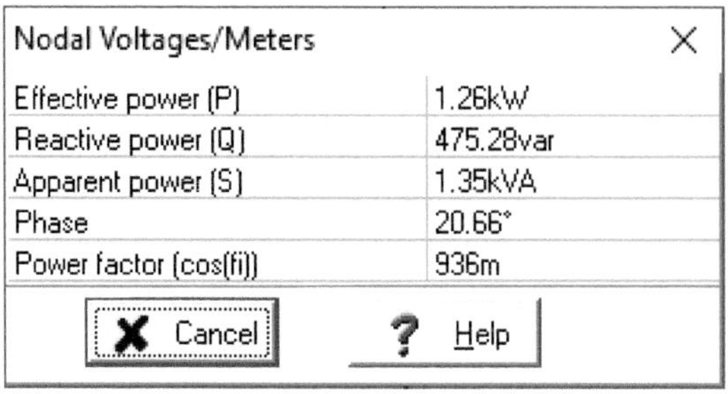

Nodal Voltages/Meters	✕
Effective power (P)	1.26kW
Reactive power (Q)	475.28var
Apparent power (S)	1.35kVA
Phase	20.66°
Power factor (cos(fi))	936m

✗ Cancel	? Help

Figure 1.134

Let's check the obtained result. Following MATLAB code calculates the average power, reactive power, apparent power and power factor.

```
clc
clear all
f=60;
Vrms=120;

R1=10;
L1=10e-3;
w=2*pi*f;
Z=R1+L1*w*j;
Irms=Vrms/abs(Z);
P=R1*Irms^2; %average(active) power
S=abs(Z)*Irms^2; %apparent power
Q=sqrt(S^2-P^2); %reactive power
pf=P/S;
disp('average power(kW) :')
disp(P/1000)
disp('reactive power(var):')
disp(Q)
disp('apparent power(kVA):')
disp(S/1000)
disp('power factor:')
disp(pf)
```

The output of MATLAB code is shown in Figure 1.135. Obtained results show that TINA-TI results are correct.

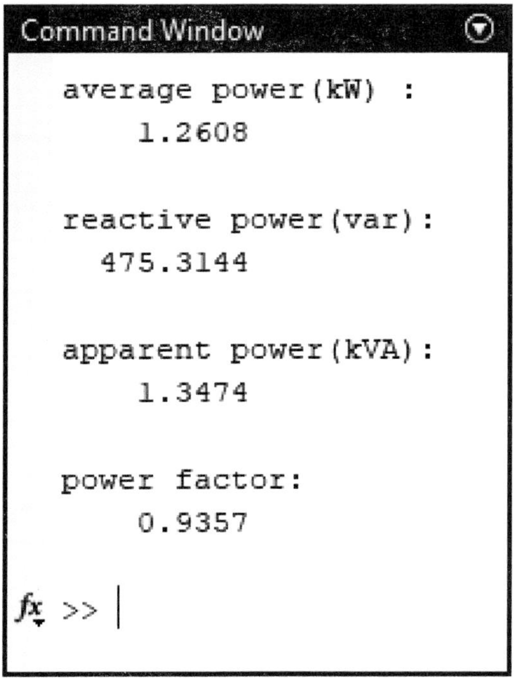

```
Command Window

    average power(kW)  :
         1.2608


    reactive power(var):
      475.3144


    apparent power(kVA):
         1.3474


    power factor:
         0.9357

fx >> |
```

Figure 1.135

1.15 Example 12: Drawing the Instantaneous Power Waveform (I)

In the previous example, we used a Power Meter block to measure the power drawn from the source. The Power Meter block can be used to observe the instantaneous power waveform as well. This example shows how you can observe instantaneous power waveform with a Power Meter block.

Open the schematic of Example 11.

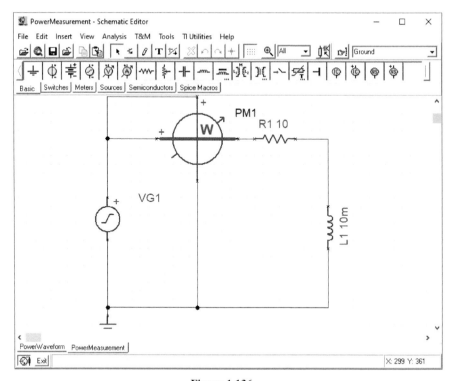

Figure 1.136

Run a transient analysis with the settings shown in Figure 1.137. The simulation result is shown in Figure 1.138. The waveform that appeared on the screen (Figure 1.138) is the graph of instantaneous power drawn from the input AC source.

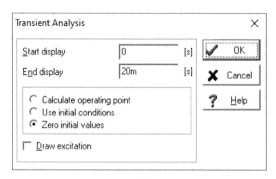

Figure 1.137

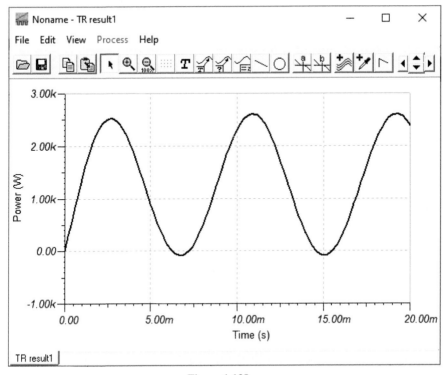

Figure 1.138

Let's measure the maximum and minimum of an obtained graph (Figure 1.139). According to Figure 1.140, maximum and minimum are 2.52 kW and −84.23 W, respectively.

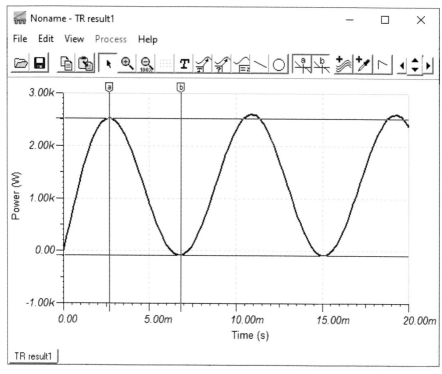

Figure 1.139

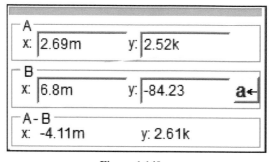

Figure 1.140

Let's check the obtained result. The following MATLAB code solves the differential equation of the circuit and draws the graph of instantaneous power drawn from the source.

```
clc
clear all

R1=10;L1=10e-3;f=60;T=1/f;w=2*pi*f;phi0=pi/4;Vm=169.7;
syms i(t) V1(t)
V1=Vm*sin(w*t+phi0);
ode=L1*diff(i,t)+R1*i==V1;
cond=i(0)==0;
iSol(t)=dsolve(ode,cond);

pLoad=simplify(Vm*sin(w*t+phi0)*iSol);
figure(1)
ezplot(pLoad,[0 0.02])
title('Instantaneous power of load')
grid minor
```

The output of MATLAB code is shown in Figure 1.141. You can compare different points of this graph with the TINA-TI results in order to ensure that both of the graphs are the same. For instance, let's compare maximum and minimum with the TINA-TI results. According to Figure 1.142, the maximum and minimum are 2527.99 W and −84.9037 W, respectively. These values are quite close to the TINA-TI result (Figure 1.140).

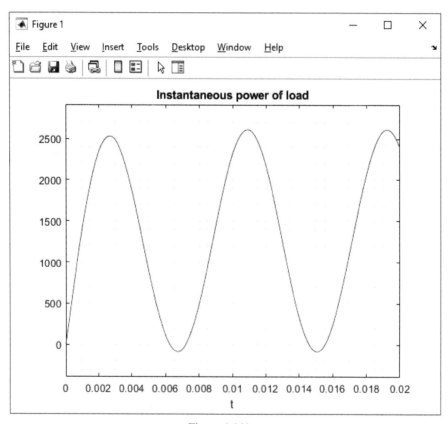

Figure 1.141

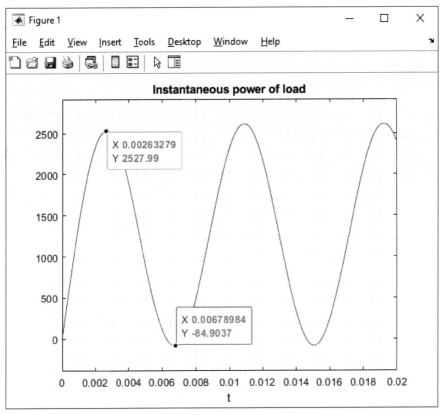

Figure 1.142

1.16 Example 13: Drawing the Instantaneous Power Waveform (II)

In the previous example, we learned one way to observe the instantaneous power waveforms. In this example, we will learn another method to generate the instantaneous power waveforms. Open the schematic of Example 12 and change it to what is shown in Figure 1.143.

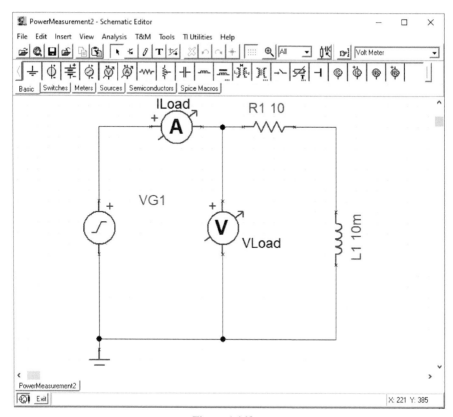

Figure 1.143

Click the Analysis> Transient (Figure 1.144).

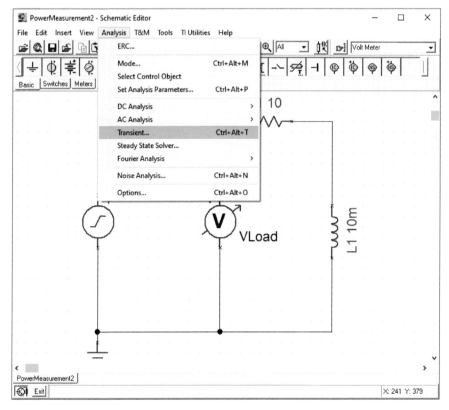

Figure 1.144

Run the transient analysis with parameters shown in Figure 1.145. The simulation result is shown in Figure 1.146.

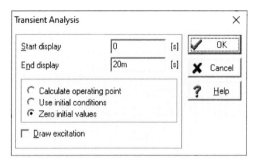

Figure 1.145

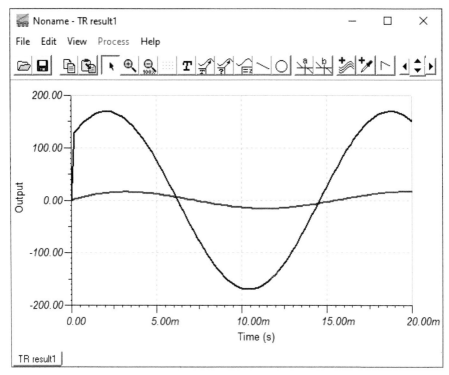

Figure 1.146

Click the Post-processor icon (Figure 1.147). This opens the post-processor window (Figure 1.148). Click the More button. The window changes to what is shown in Figure 1.149.

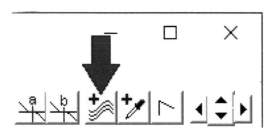

Figure 1.147

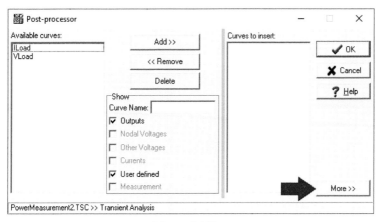

Figure 1.148

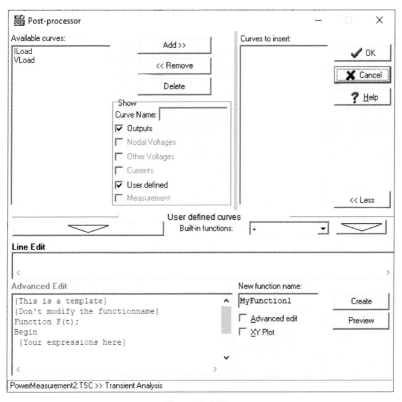

Figure 1.149

Click the VLoad and down arrow button. This adds the VLoad(t) to Line Edit box (Figure 1.150).

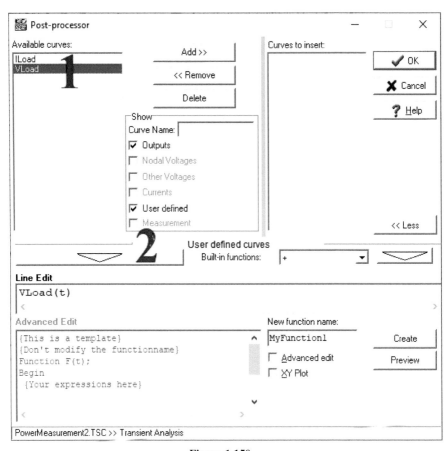

Figure 1.150

Click on the Line Edit box and enter a * into it (Figure 1.151).

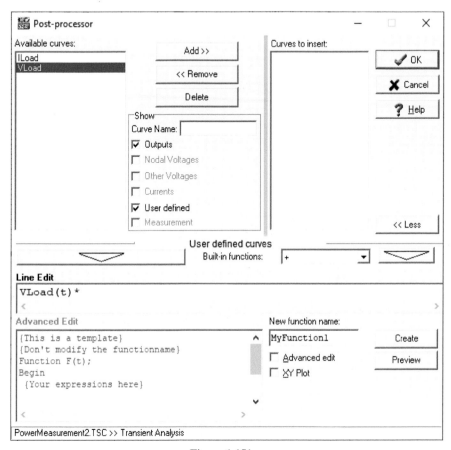

Figure 1.151

Click the ILoad and down arrow button. This adds the ILoad(t) to Line Edit box (Figure 1.152).

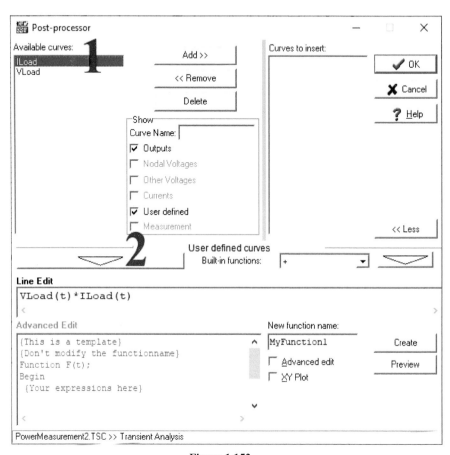

Figure 1.152

Enter a name for the entered function. Then click the Create button.

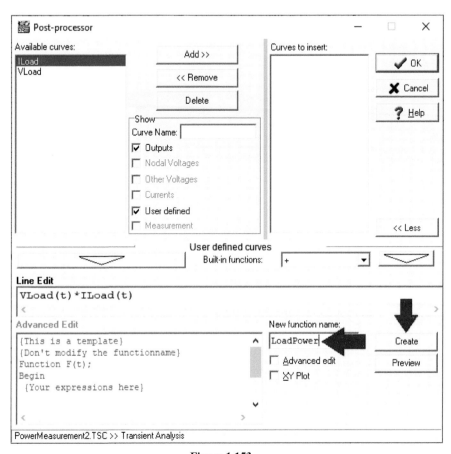

Figure 1.153

After clicking the Create button, entered function name is added to the 'Curves to insert' list (Figure 1.154). Click the OK button to add the graph of an entered function to the output window (Figure 1.155).

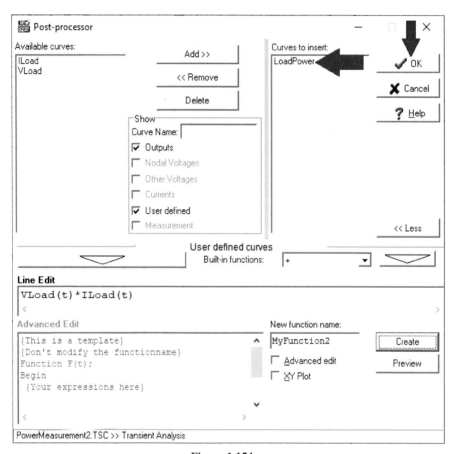

Figure 1.154

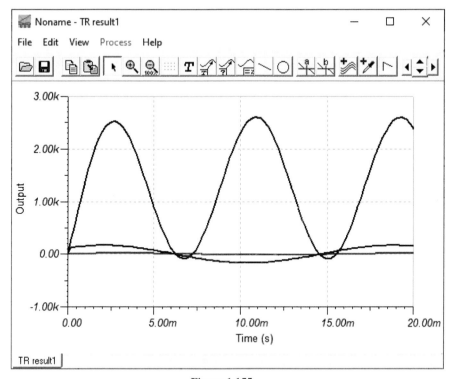

Figure 1.155

Figure 1.155 contains three waveforms: circuit current, circuit voltage and product of these two waveforms (=instantaneous power). Let's remove the circuit current and voltage waveforms from the output window. Click the View> Show/Hide curves (Figure 1.156). This opens the Show/hide curves window (Figure 1.157). Remove the checkmark behind Vload and ILoad boxes (Figure 1.158) and click the Close button. After clicking the Close button, only the graph of instantaneous load power appears on the screen (Figure 1.159).

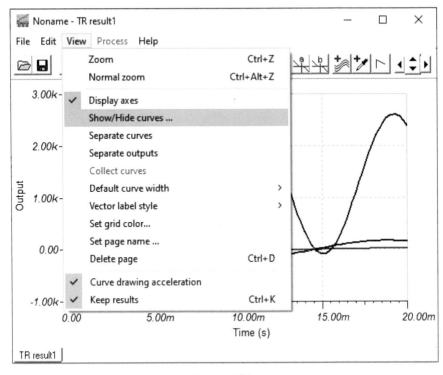

Figure 1.156

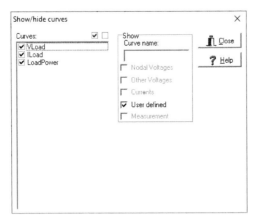

Figure 1.157

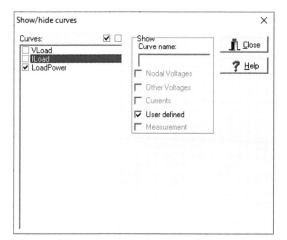

Figure 1.158

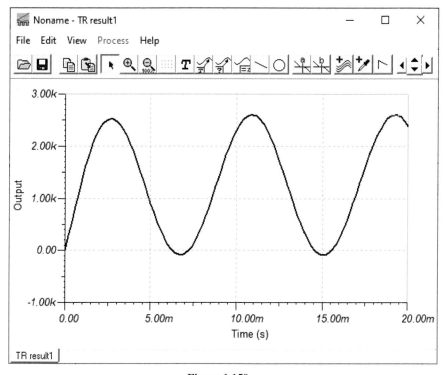

Figure 1.159

Let's change the label of the vertical axis in Figure 1.159 to 'Load Power (W)'. Double click on the vertical axis in Figure 1.159. This opens the Set Axis window (Figure 1.160).

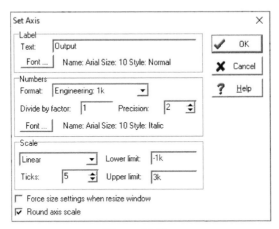

Figure 1.160

Enter 'Load Power (W)' into the Text box (Figure 1.161) and click the OK button. Now, the vertical axis has a 'Load Power (W)' label (Figure 1.162).

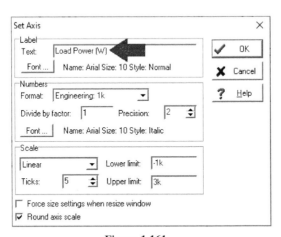

Figure 1.161

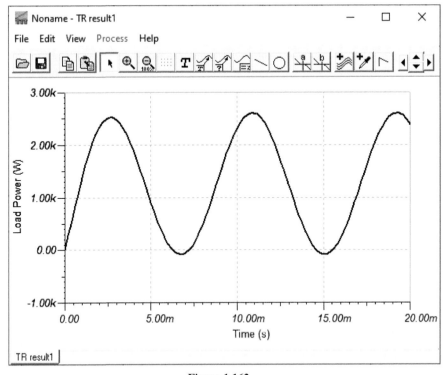

Figure 1.162

1.17 Example 14: Ohm Meter block (I)

The Ohm meter block (Figure 1.163) can be used to measure the magnitude of the impedance of a network at a desired frequency. Let's study an example. Draw the schematic shown in Figure 1.164. Settings of Ohm meter block ZM1 are shown in Figure 1.165. The block ZM1 measures the magnitude of the impedance of the network shown in Figure 1.164 at 60 Hz.

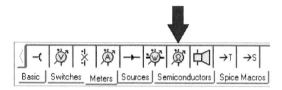

Figure 1.163

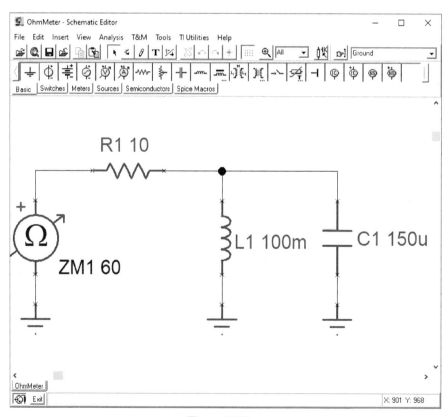

Figure 1.164

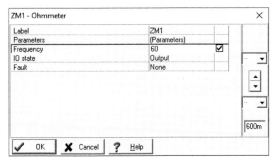

Figure 1.165

Click the Analysis> AC Analysis> Calculate nodal voltages (Figure 1.166). The simulation result is shown in Figure 1.167.

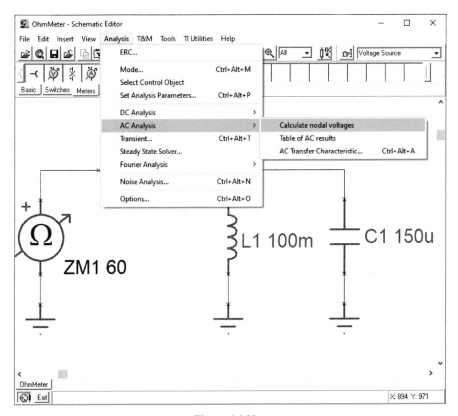

Figure 1.166

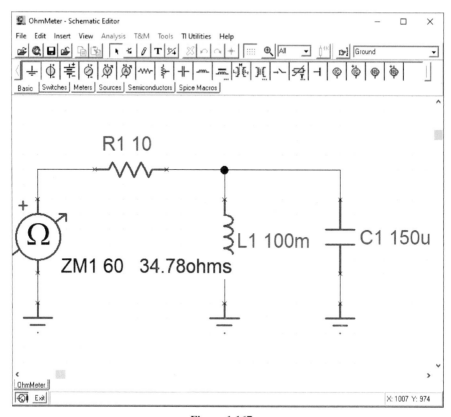

Figure 1.167

Let's check the obtained result. The MATLAB code shown in Figure 1.168 calculates the impedance of the network at 60 Hz. The impedance of the network at 60 Hz is $10 - 33.308j\ \Omega$. The magnitude of this impedance is $|10 - 33.308j\,| = \sqrt{10^2 + 33.308^2} = 34.7767\ \Omega$. The number which is shown by TINA-TI is quite close to this number.

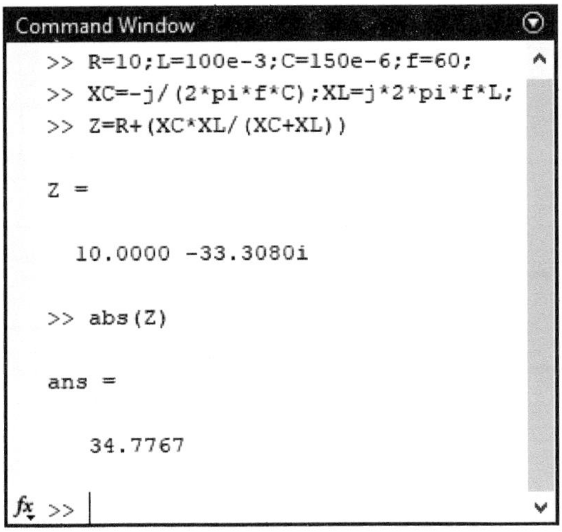

```
Command Window
>> R=10;L=100e-3;C=150e-6;f=60;
>> XC=-j/(2*pi*f*C);XL=j*2*pi*f*L;
>> Z=R+(XC*XL/(XC+XL))

Z =

   10.0000 -33.3080i

>> abs(Z)

ans =

   34.7767

fx >>
```

Figure 1.168

1.18 Example 15: Ohm Meter Block (II)

The Ohm meter block can be used to measure the impedance at 0 Hz (DC) as well. Let's study an example. Draw the schematic shown in Figure 1.169. Settings of the ZM1 block are shown in Figure 1.170. Note that the frequency that is entered into the Frequency box has no importance when you want to measure the DC impedance of the network.

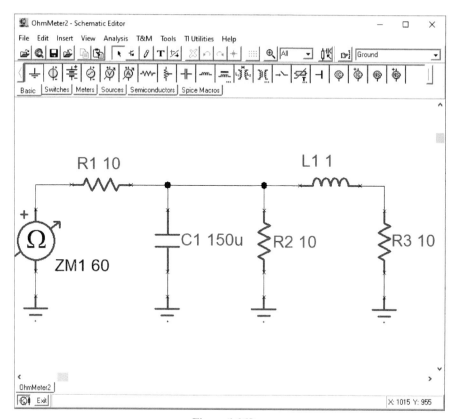

Figure 1.169

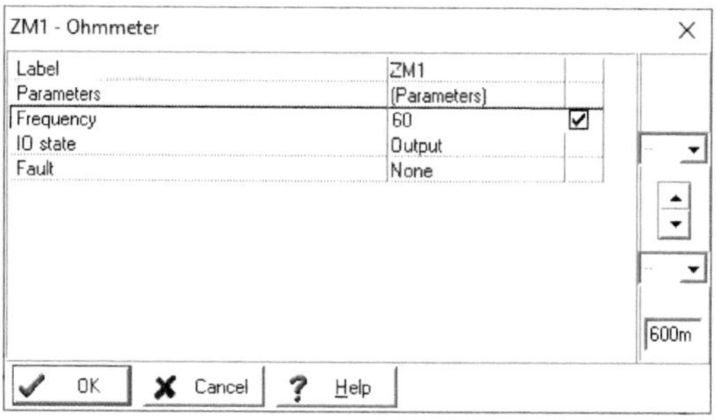

Figure 1.170

Click the Analysis> DC Analysis> Calculate nodal voltages (Figure 1.171). The simulation result is shown in Figure 1.172. According to Figure 1.172, the DC impedance (resistance) of this network is 15 Ω.

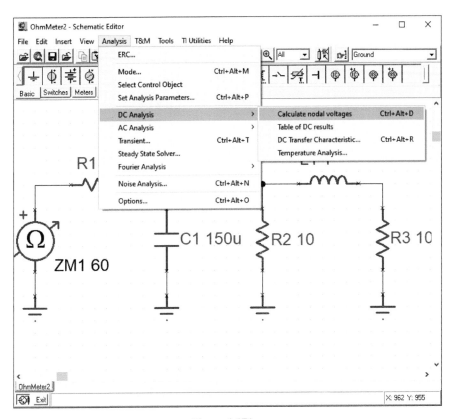

Figure 1.171

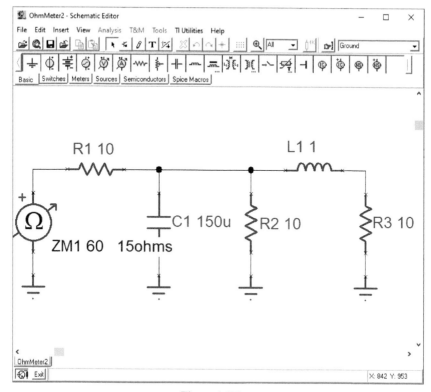

Figure 1.172

Let's check the obtained result. The DC equivalent circuit is shown in Figure 1.173. Remember that capacitors act like open circuits and inductors act like a short circuits in steady-state DC conditions. The equivalent resistance of Figure 1.173 is $10 + \frac{10 \times 10}{10+10} = 15\ \Omega$.

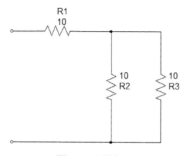

Figure 1.173

1.19 Example 16: Thevenin Equivalent Circuit

In this example, we want to find the Thevenin equivalent circuit with respect to the terminals 'a' and 'b' for the circuit shown in Figure 1.174.

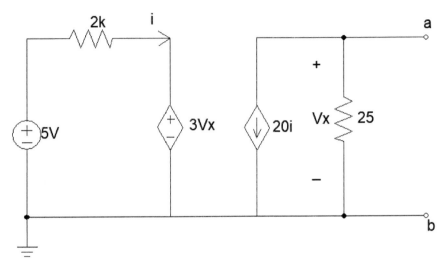

Figure 1.174

Let's start. Click the Controlled Sources icon (Figure 1.175). After clicking, the menu shown in Figure 1.176 appears on the screen. The circuit in Figure 1.174 requires a Voltage Controlled Voltage Source (VCVS) and a Current Controlled Current Source (CCCS) block. Add a VCVS block and a CCCS block to the schematic and draw what is shown in Figure 1.177. Figure 1.177 is the TINA-TI equivalent of Figure 1.174.

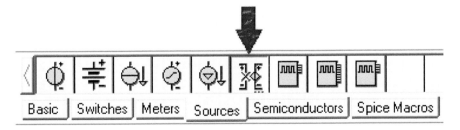

Figure 1.175

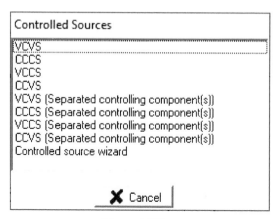

Figure 1.176

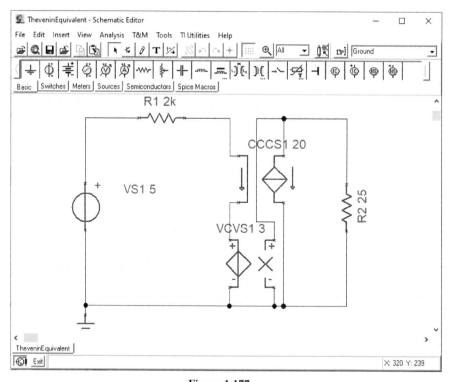

Figure 1.177

Settings of VCVS and CCCS blocks are shown in Figure 1.178 and Figure 1.179, respectively.

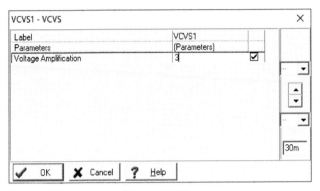

Figure 1.178

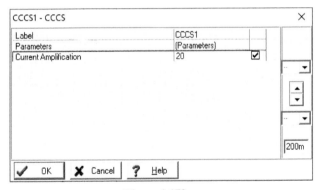

Figure 1.179

Our schematic is ready. Add a Voltmeter block to the schematic (Figure 1.180). This Voltmeter measures the Thevenin voltage for us. Remember that Thevenin voltage (V_{TH}) is the open-circuit voltage between the terminals.

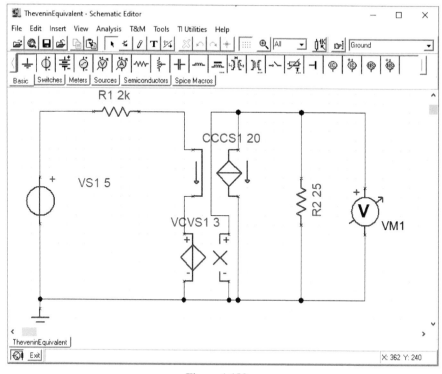

Figure 1.180

Click the Analysis> DC Analysis> Calculate nodal voltages (Figure 1.181). Simulation result is shown in Figure 1.182. According to Figure 1.182, $V_{AB} = V_A - V_B = -5V$. So, $V_{TH} = V_{AB} = V_A - V_B = -5V$.

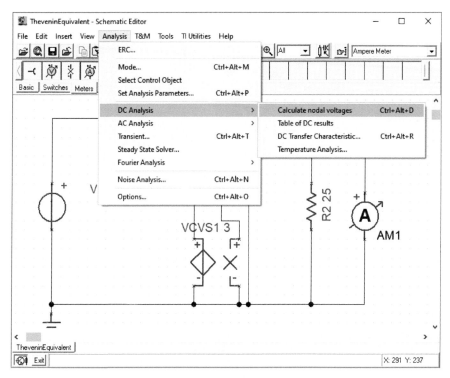

Figure 1.181

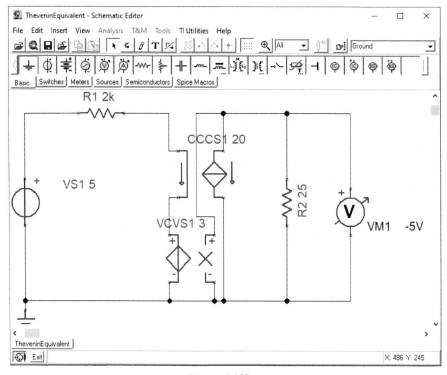

Figure 1.182

It is time to measure the Thevenin resistance (R_{TH}). We can measure the Thevenin resistance with the aid of a short circuit current (I_{SC}). Remember that $R_{TH} = \frac{V_{TH}}{I_{SC}}$. The I_{SC} is measured with the aid of the schematic shown in Figure 1.183.

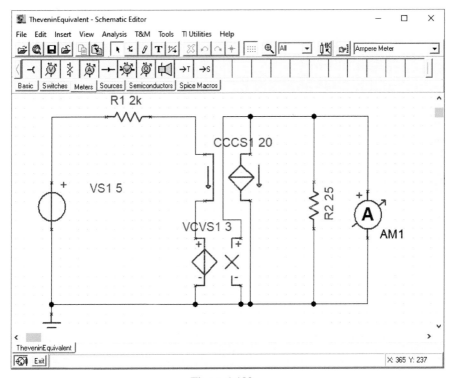

Figure 1.183

Click the Analysis> DC Analysis> Calculate nodal voltages (Figure 1.181). The simulation result is shown in Figure 1.184. According to Figure 1.184, $I_{SC} = -50\ mA$. So, $R_{TH} = \frac{V_{TH}}{I_{SC}} = \frac{-5\ V}{50\ mA} = 100\ \Omega$. The Thevenin equivalent circuit for a given circuit is shown in Figure 1.185.

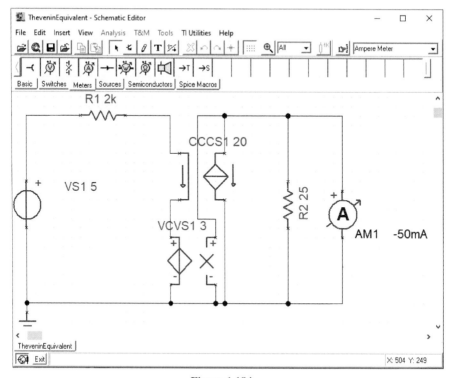

Figure 1.184

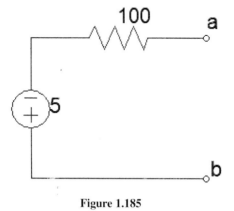

Figure 1.185

1.20 Example 17: Measurement of Thevenin Resistance with the Aid of Ohm Meter Block

In the previous example, we introduced one method to measure the Thevenin resistance. The Thevenin resistance can be measured with the aid of Ohm meter block as well. In this example, we will measure the Thevenin resistance of Example 16 with the aid of the Ohm meter block. Draw the schematic shown in Figure 1.186. Note that the independent voltage source is killed. Remember that in order to kill an independent voltage source, it must be replaced with a short circuit and in order to kill an independent current source, it must be replaced with an open circuit.

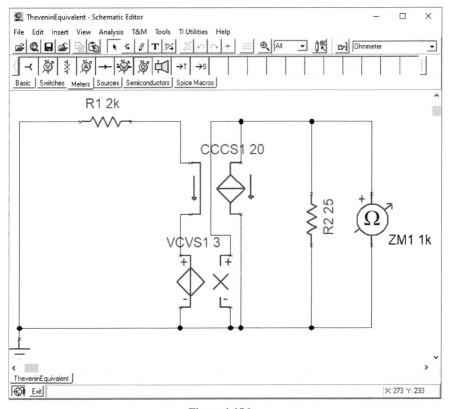

Figure 1.186

Settings of the Ohm meter block are shown in Figure 1.187. Note that the value entered into the Frequency box has no effect on our measurement.

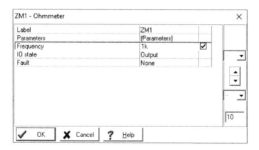

Figure 1.187

Click the Analysis> DC Analysis> Calculate nodal voltages (Figure 1.188). The simulation result is shown in Figure 1.189. According to Figure 1.189, $R_{TH} = 100\ \Omega$. This is the same as the result we obtained in Example 16.

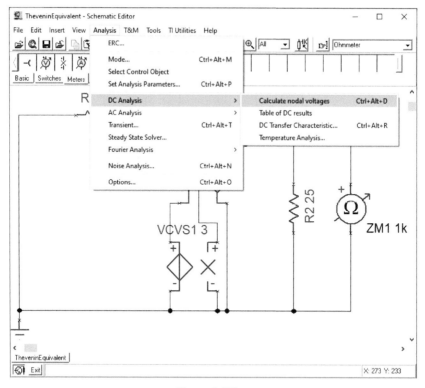

Figure 1.188

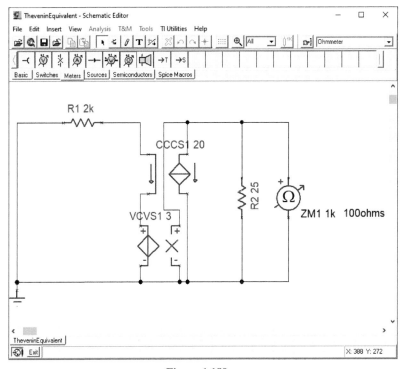

Figure 1.189

1.21 Example 18: Current Controlled Voltage Source

We used current-controlled current source and VCCS blocks in Example 16. This example shows how to simulate circuits that contain CCVS. Consider the circuit shown in Figure 1.190. We want to measure the voltage of node 'out'. From basic circuit theory, we expect the voltage of node 'out' to be 10 V.

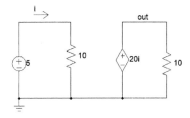

Figure 1.190

Draw the schematic shown in Figure 1.191. This schematic uses a CCVS block (Figure 1.192). Settings of CCVS block CCVS1 are shown in Figure 1.193.

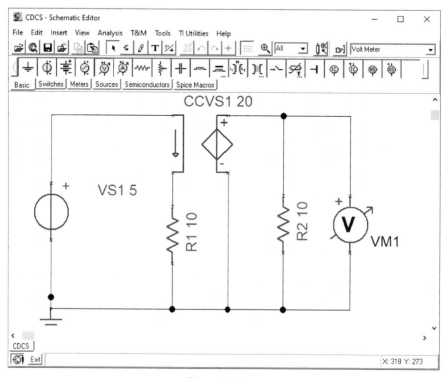

Figure 1.191

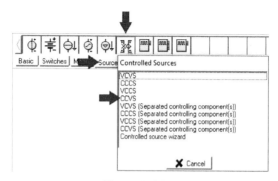

Figure 1.192

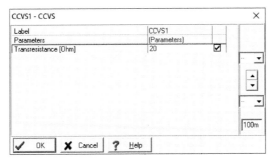

Figure 1.193

Click the Analysis> DC Analysis> Calculate nodal voltages (Figure 1.194). The simulation result is shown in Figure 1.195. The voltage is 10 V as expected.

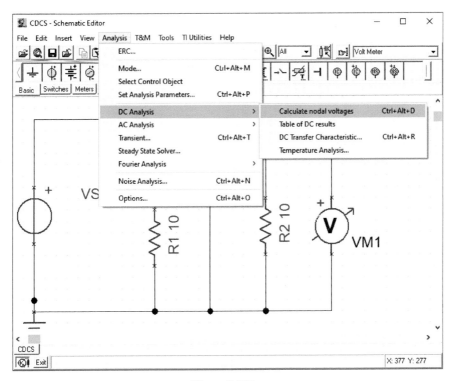

Figure 1.194

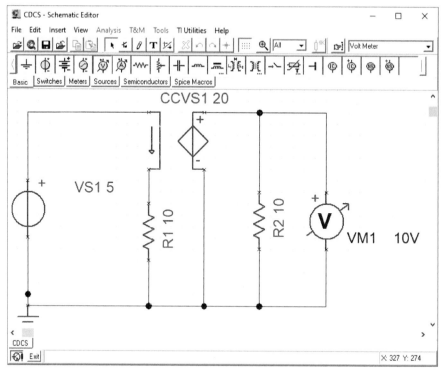

Figure 1.195

1.22 Example 19: Voltage Controlled Current Sources Block

In this example, we see how to simulate VCCS in TINA-TI. Consider the schematic shown in Figure 1.196. This circuit contains a VCCS. We want to measure the value of V_x. From basic circuit theory, $V_x + 2V_x - 0.3V_x = 5$ or $V_x = 1.8518\,V$.

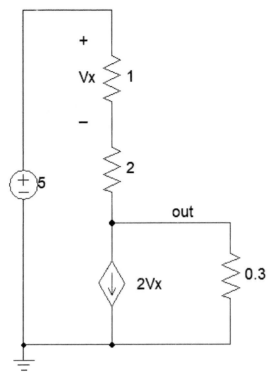

Figure 1.196

Draw the schematic shown in Figure 1.197. This schematic uses a VCCS block (Figure 1.198). Settings of VCCS block are shown in Figure 1.199.

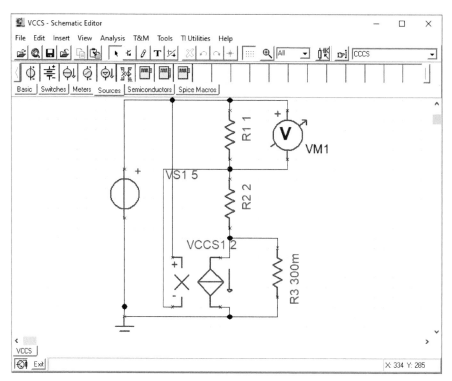

Figure 1.197

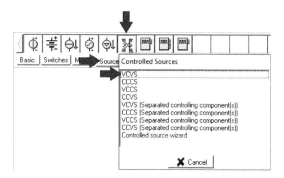

Figure 1.198

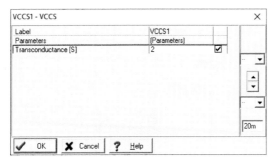

Figure 1.199

Click the Analysis> DC Analysis> Calculate nodal voltages (Figure 1.200). The simulation result is shown in Figure 1.201. The voltage is 1.85 V as expected.

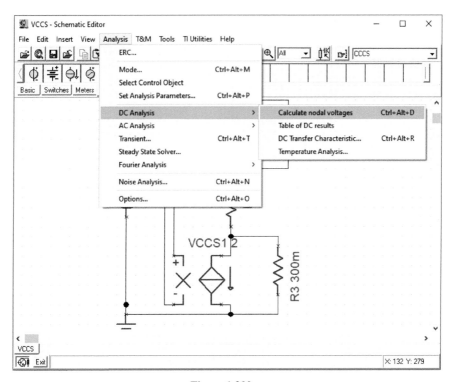

Figure 1.200

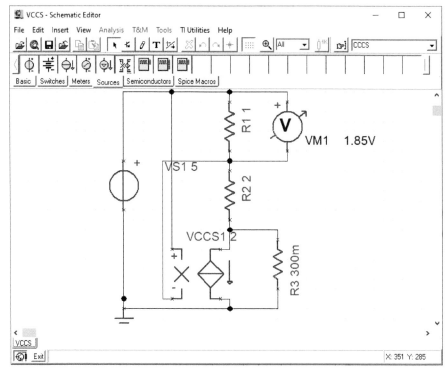

Figure 1.201

1.23 Example 20: Switch Block

In this example, we want to simulate the circuit shown in Figure 1.202. The initial voltage of the capacitor is 0 V. Switch S1 is closed at t= 0 and is opened at t=3 s. Switch S2 is open at [0, 3 s] interval and is closed at t=3 s.

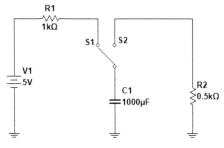

Figure 1.202

The status of switches is shown in Figure 1.203. We want to observe the capacitor voltage for [0, 10 s] time interval.

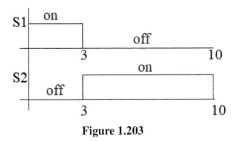

Figure 1.203

Draw the schematic shown in Figure 1.204. This schematic uses a Time-Controlled Switch block (Figure 1.205).

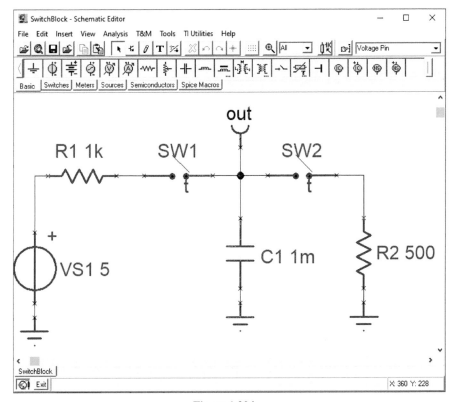

Figure 1.204

Figure 1.205

Settings of SW1 and SW2 are shown in Figure 1.206 and Figure 1.207, respectively. The 't On (s)' and 't Off (s)' boxes are filled according to Figure 1.203.

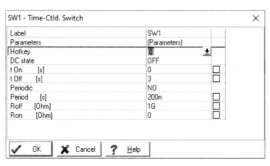

Figure 1.206

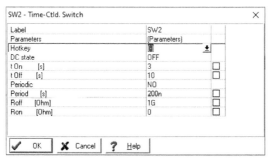

Figure 1.207

It is recommended to click the Help button in Figure 1.206 or Figure 1.207 and study the description of each parameter carefully (Figure 1.208).

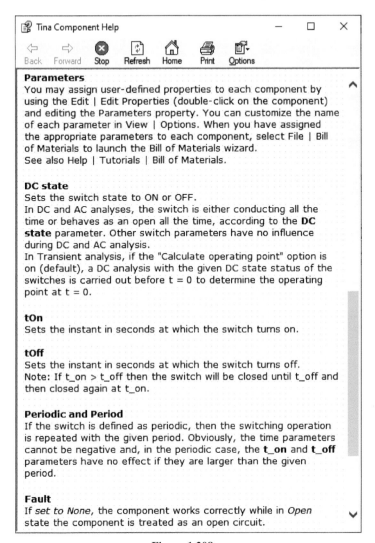

Figure 1.208

Run a transient analysis with the parameters shown in Figure 1.209. The simulation result is shown in Figure 1.210. The capacitor is charged during the [0, 3 s] interval and it is discharged during the [3 s, 10 s] interval.

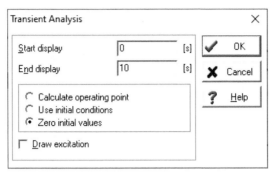

Figure 1.209

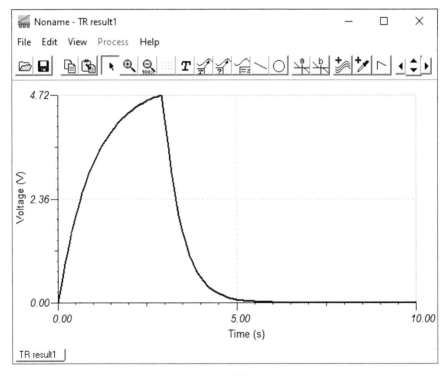

Figure 1.210

1.24 Example 21: Three Phase Source

A three-phase source can be simulated with the aid of three Voltage Generator blocks (Figure 1.211). In this example, we will learn how to simulate a three-phase source.

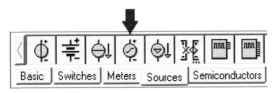

Figure 1.211

Delta (Δ) connected three-phase sources can be simulated with the aid of schematic shown in Figure 1.212. Small resistors R1, R2 and R3 are put for convergence issues. Settings of VG1, VG2 and VG3 are shown in Figure 1.213, Figure 1.214 and Figure 1.215, respectively. Note that 120° of phase difference exists between the voltage generators.

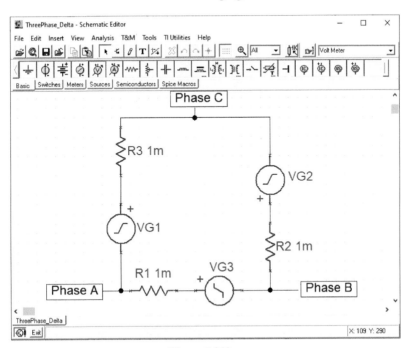

Figure 1.212

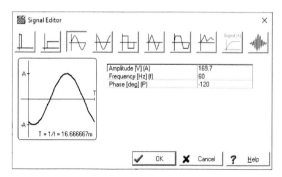

Figure 1.213

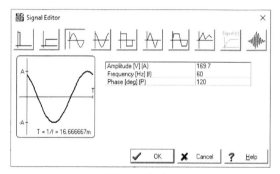

Figure 1.214

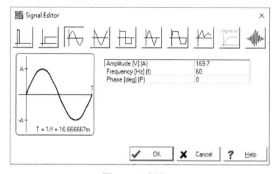

Figure 1.215

Star (Y) connected three-phase sources can be simulated with the aid of schematic shown in Figure 1.216. Settings of VG1, VG2 and VG3 are shown in Figure 1.213, Figure 1.214 and Figure 1.215, respectively.

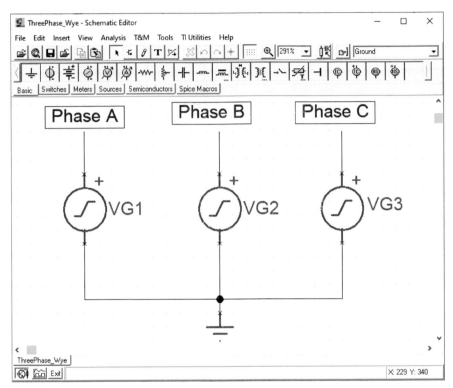

Figure 1.216

1.25 Example 22: Jumper Block

Use jumpers (Figure 1.217) to connect distant components and nodes into one net without cluttering the schematic with explicit wires.

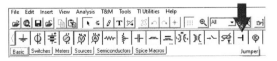

Figure 1.217

Let's study an example. Consider the schematic shown in Figure 1.218. Settings of VG1, VG2 and VG3 are shown in Figure 1.213, Figure 1.214 and Figure 1.215, respectively.

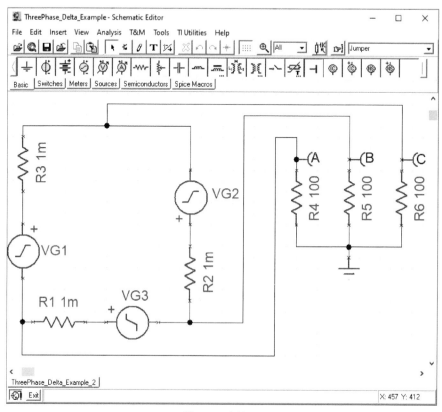

Figure 1.218

The schematic shown in Figure 1.218 can be redrawn with the aid of jumper block as shown in Figure 1.219. Both of the schematics in Figure 1.218 and Figure 1.219 are the same from the TINA-TI point of view. However, schematically shown in Figure 1.219 is more understandable for the human user.

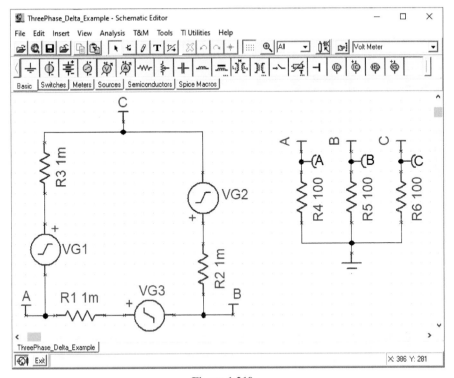

Figure 1.219

Let's simulate the circuit and see the voltages of nodes 'A', 'B' and 'C'. Click the Analysis> Transient (Figure 1.220). Do the transient analysis settings similar to Figure 1.121 and run the simulation. The simulation result is shown in Figure 1.222.

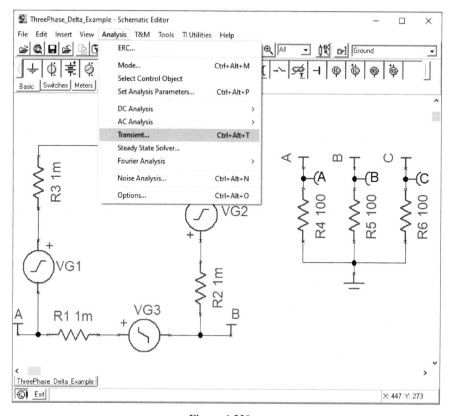

Figure 1.220

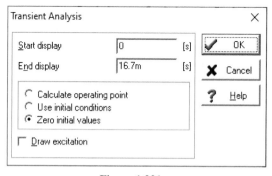

Figure 1.221

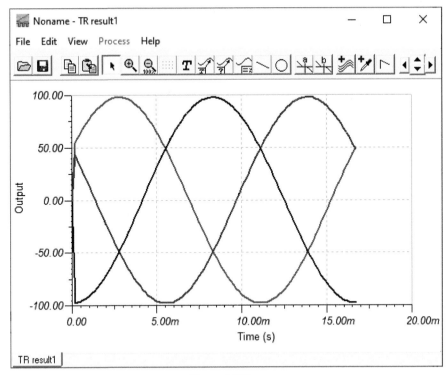

Figure 1.222

Click the legend icon (Figure 1.223) to see which curve belongs to which node (Figure 1.224).

Figure 1.223

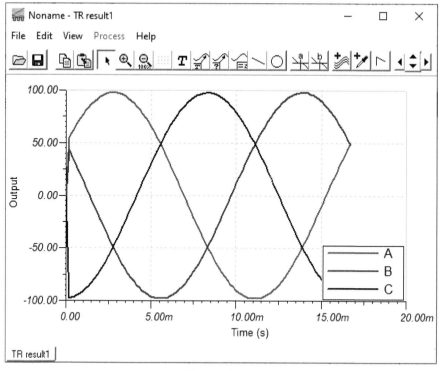

Figure 1.224

Let's check the obtained result. Following equations can be written according to Figure 1.218.

$$\frac{v_A}{R} + \frac{v_A - v_{G3}}{R} + \frac{v_A - v_{G1}}{R} = 0 \Rightarrow v_A = \frac{v_{G3} - v_{G1}}{3}$$

$$\frac{v_B}{R} + \frac{v_B + v_{G3}}{R} + \frac{v_B - v_{G2}}{R} = 0 \Rightarrow v_B = \frac{v_{G2} - v_{G3}}{3}$$

$$\frac{v_C}{R} + \frac{v_C + v_{G2}}{R} + \frac{v_C - v_{G1}}{R} = 0 \Rightarrow v_C = \frac{v_{G1} - v_{G2}}{3}$$

The MATLAB code shown in Figure 1.225 draws the graph of these equations. The output of the code is shown in Figure 1.226. The obtained result is the same as the TINA-TI result shown in Figure 1.222.

```
Command Window                              ⊙
  >> syms t
  >> w=2*pi*60;
  >> VG1=169.7*sin(w*t-(2*pi/3));
  >> VG2=169.7*sin(w*t+(2*pi/3));
  >> VG3=169.7*sin(w*t);
  >> ezplot((VG3-VG1)/3,[0,16.7e-3]), hold on
  >> ezplot((VG2-VG3)/3,[0,16.7e-3])
  >> ezplot((VG1-VG2)/3,[0,16.7e-3])
fx >> |
```

Figure 1.225

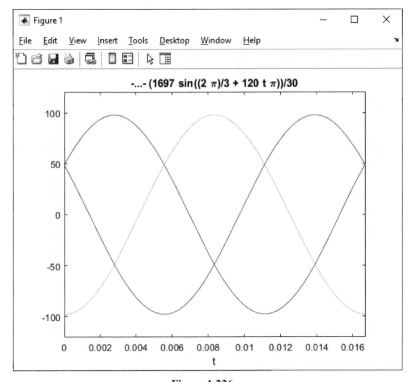

Figure 1.226

1.26 Example 23: Coupled Inductors

Coupled Inductors block (Figure 1.227) can be used to simulate coupled inductor coupled inductors.

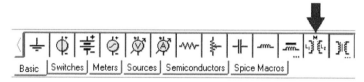

Figure 1.227

Let's study an example. Consider the circuit shown in Figure 1.228. Vin is a step voltage and M is the mutual inductance between L1 and L2. The coupling coefficient between the two coils is $k = \dfrac{M}{\sqrt{L_1 L_2}} = \dfrac{0.9m}{\sqrt{1m \times 1.1m}} = 0.8581$.

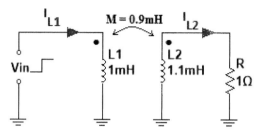

Figure 1.228

From basic circuit theory,

$$\begin{cases} L_1 \frac{di_{L1}}{dt} - M \frac{di_{L2}}{dt} = V_{in}(t) \\ Ri_{L2} + L_2 \frac{di_{L2}}{dt} - M \frac{di_{L1}}{dt} = 0 \end{cases}$$

After taking the Laplace transform of both side:

$$\begin{bmatrix} L_1 s & -Ms \\ -Ms & R + L_2 s \end{bmatrix} \times \begin{bmatrix} I_{L1}(s) \\ I_{L2}(s) \end{bmatrix} = \begin{bmatrix} V_{in}(s) \\ 0 \end{bmatrix}$$

So,

$$\begin{bmatrix} I_{L1}(s) \\ I_{L2}(s) \end{bmatrix} = \begin{bmatrix} L_1 s & -Ms \\ -Ms & R + L_2 s \end{bmatrix}^{-1} \times \begin{bmatrix} V_{in}(s) \\ 0 \end{bmatrix}$$

$V_{in}(s) = \frac{1}{s}$, so

$$\begin{bmatrix} I_{L1}(s) \\ I_{L2}(s) \end{bmatrix} = \begin{bmatrix} \frac{(11s+10000) \times 10000}{s^2 \times (29s+100000)} \\ \frac{90000}{s(29s+100000)} \end{bmatrix}$$

You can use the following MATLAB commands to see the time domain graph of I_{L1} and I_{L2}. Outputs of the code are shown in Figures 1.229 and 1.230.

```
s=tf('s');
I1=(11*s+10000)*10000/s/(29*s+100000);
I2=90000/(29*s+100000);
figure(1)
step(I1,[0:0.003/100:0.003]),grid on
figure(2)
step(I2), grid on
```

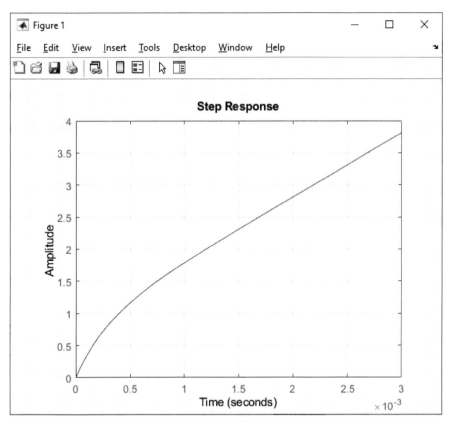

Figure 1.229

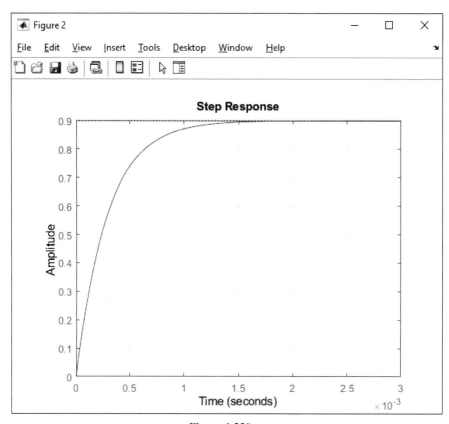

Figure 1.230

Let's solve this problem with TINA-TI. Draw the schematic shown in Figure 1.231. IL1 and IL2 are used to measure the current I_{L1} and I_{L2}, respectively.

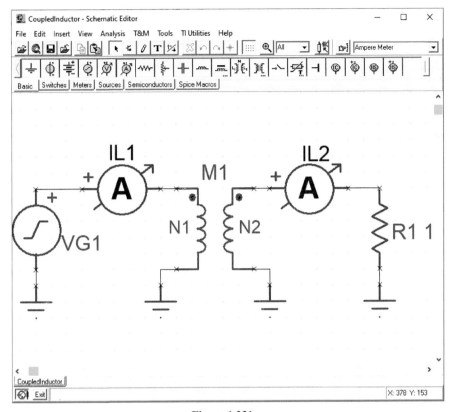

Figure 1.231

Settings of Coupled Inductors block M1 is shown in Figure 1.232.

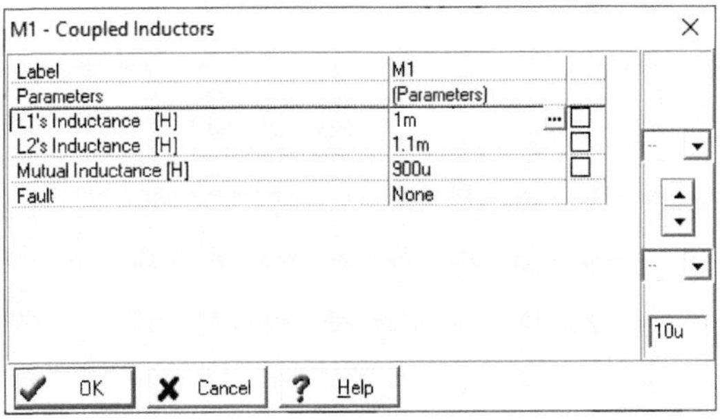

Figure 1.232

Settings of Voltage Generator block VG1 is shown in Figures 1.233 and 1.234.

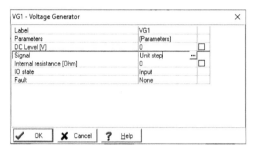

Figure 1.233

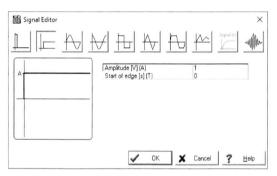

Figure 1.234

Run a transient analysis with the parameters shown in Figure 1.235. The simulation result is shown in Figure 1.236.

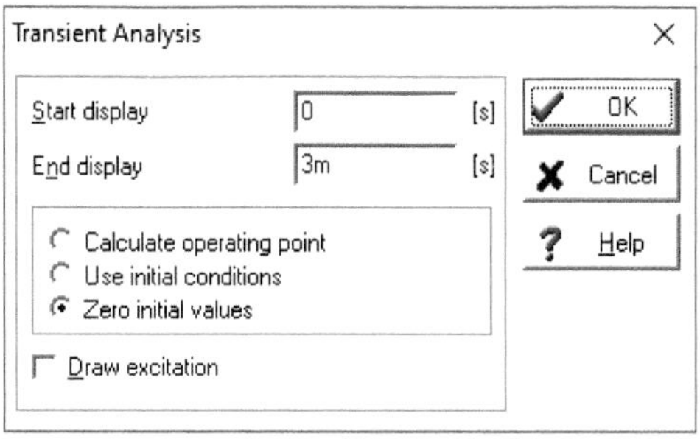

Figure 1.235

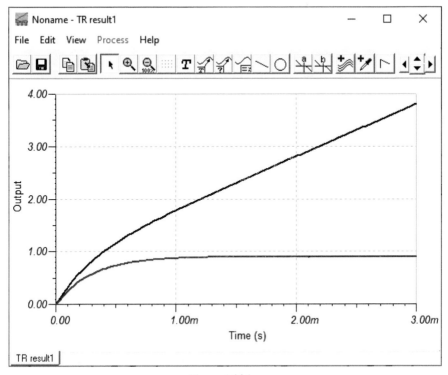

Figure 1.236

Click the View> Separate curves (Figure 1.237) to see the graphs on two separate coordinate systems (Figure 1.238). You can use the cursors in order to ensure that IL1 and IL2 graphs in Figure 1.238 are the same as the graphs in Figures 1.229 and 1.230.

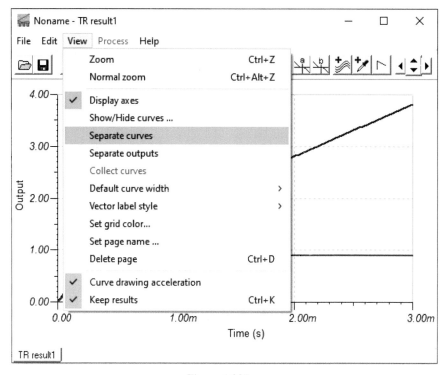

Figure 1.237

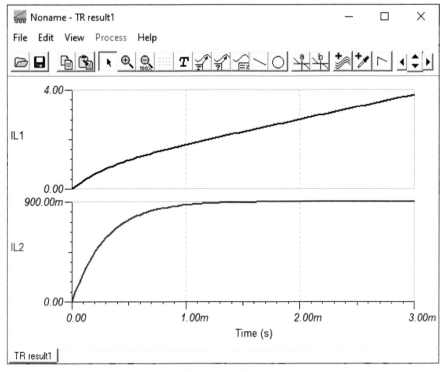

Figure 1.238

1.27 Example 24: Transformer

In this example, we want to measure the RMS of the current drawn from voltage source V1 in Figure 1.239.

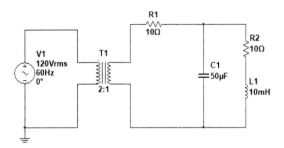

Figure 1.239

The load connected to the secondary of the transformer is shown in Figure 1.240. The impedance of this load is calculated with the aid of MATLAB commands shown in Figure 1.241.

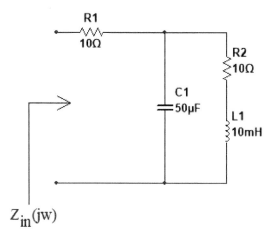

Figure 1.240

```
Command Window
>> f=60;w=2*pi*f;R1=10;C1=50e-6;R2=10;L1=10e-3;
>> Xc=-j/(w*C1);
>> XL=j*w*L1;
>> Z2=R2+XL;
>> Z=R1+Xc*Z2/(Xc+Z2)

Z =

   21.1302 + 1.7998i

fx >>
```

Figure 1.241

The impedance seen from primary side is $n^2 \times Z_{in}(j\omega) = 2^2 \times (21.1302 + 1.7998j) = 84.5208 + 7.1992j$. Therefore, RMS of current drawn from input AC source is $\left|\dfrac{120}{84.5208+7.1992j}\right| = 1.4146\ A$.

The circuit in Figure 1.239 has a transformer. In TINA-TI, the transformer can be simulated with the aid of the Transformers icon (Figure 1.242). After clicking the Transformers icon, the list shown in Figure 1.243 appears on the screen. Figure 1.244 shows more information about each of the options in this list.

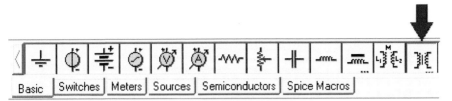

Figure 1.242

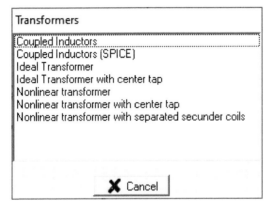

Figure 1.243

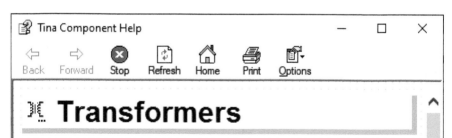

Description

There are many ways to define transformers in TINA:

A **Ideal Transformer** is an ideal transformer with only one parameter, its voltage transfer ratio ($N=V_2/V_1$); that is, the ratio of the voltages of the secondary (2) and primary (1) sides of the transformer. When the transformer symbol is placed in its default orientation (no rotation or mirroring), the secondary winding on the right hand side. It is a good idea to place a label on the secondary winding so that if you ever mirror or rotate the transformer, the label will continue to indicate the secondary winding.

Ideal Transformer with a center tap: This ideal transformer has a center tap in the secondary, otherwise it is the same as the classic ideal transformer. The voltage ratio considers the full secondary coil.

Nonlinear Transformer: For this model of a real transformer, you can select a core and define other parameters of the transformer model.

Nonlinear Transformer with center tap: This is the same as the nonlinear transformer, only with a tap in the middle of the secondary coil.

Nonlinear Transformer with separated secondary coils: Same as the nonlinear transformer, but with two isolated secondary coils.

Figure 1.244

Draw the schematic shown in Figure 1.245. Note that TR1 is an 'Ideal Transformer' block.

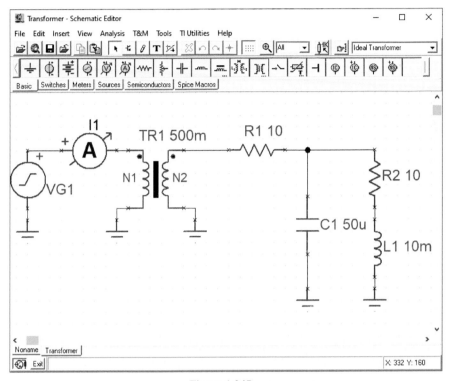

Figure 1.245

Settings of VG1 and TR1 blocks are shown in Figure 1.246 and Figure 1.247, respectively. Note that the Ratio box in Figure 1.247 takes the $\frac{N_s}{N_p}$ where N_s and N_p show the number of turns on the secondary side and the number of turns on the primary side, respectively. According to Figure 1.239, $\frac{N_s}{N_p} = 0.5$.

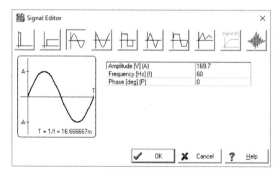

Figure 1.246

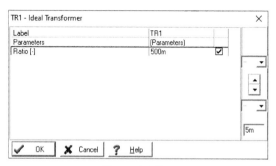

Figure 1.247

Run a transient analysis with the parameters shown in Figure 1.248. The simulation result is shown in Figure 1.249. According to Figure 1.249, the peak of the current is 2 A. So, its RMS is $\frac{2}{\sqrt{2}} = 1.41\ A$.

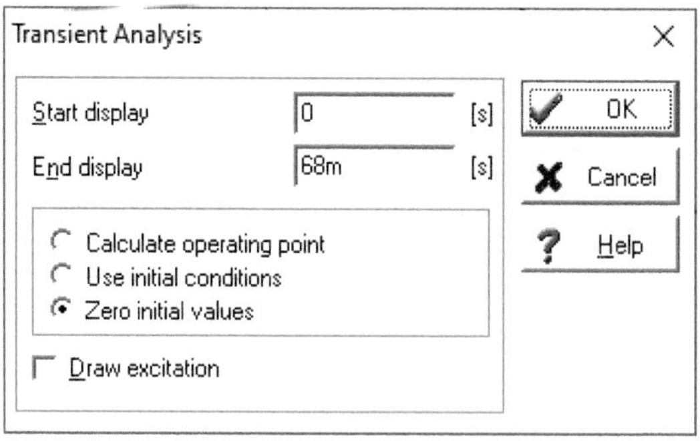

Figure 1.248

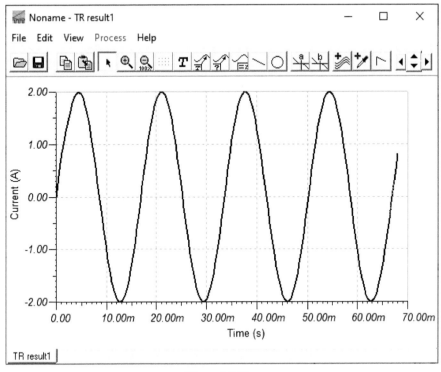

Figure 1.249

1.28 Example 25: Unit Impulse Response of Electric Circuits

TINA-TI can be used to calculate the unit impulse response of electric circuits. In this example, we want to obtain the impulse response of an RLC circuit.

Let's start. Draw the schematic shown in Figure 1.250.

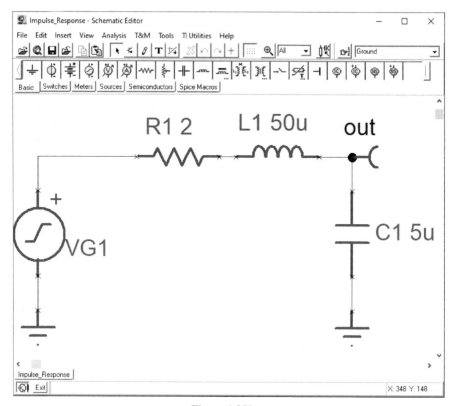

Figure 1.250

Settings of the block VG1 is shown in Figure 1.251 and Figure 1.252.

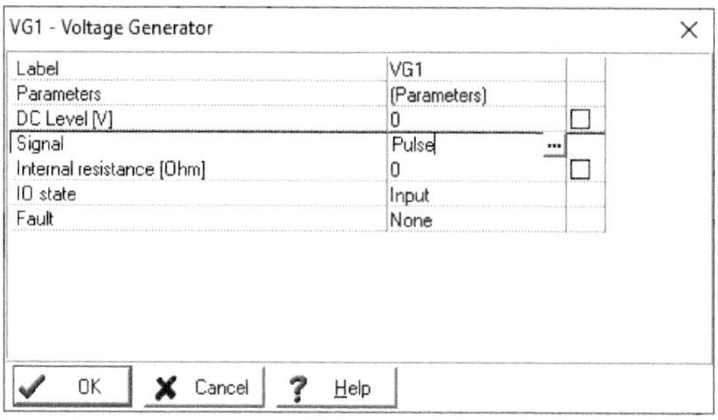

Figure 1.251

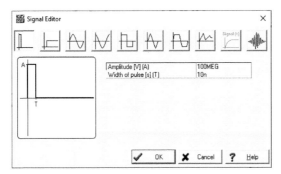

Figure 1.252

The settings shown in Figure 1.252 generate the waveform shown in Figure 1.253 ($A = 10^8$ and $T = 10$ ns). Note that the integral of the waveform shown in Figure 1.253 is $\int_{-\infty}^{+\infty} f(t)dt = A \times T = 1$.

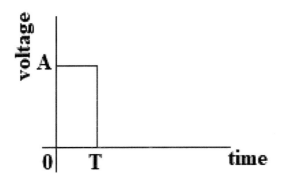

Figure 1.253

Run a transient analysis with the parameters shown in Figure 1.254. The simulation result is shown in Figure 1.255.

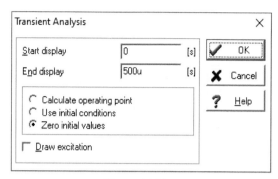

Figure 1.254

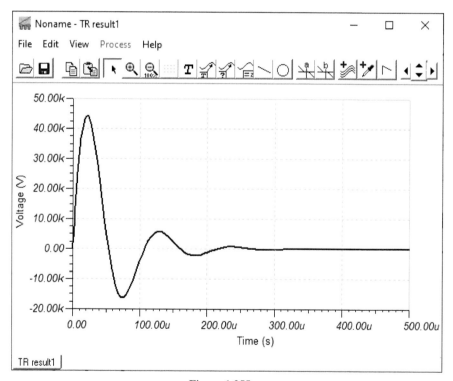

Figure 1.255

Let's check the obtained result. The MATLAB code shown in Figure 1.256 draws the unit impulse response of the circuit. Note that the transfer function of the RLC circuit in Figure 1.250 is $\frac{V_{C_1}(s)}{V_1(s)}$ =

$\dfrac{\frac{1}{C_1 s}}{R_1 + L_1 s + \frac{1}{C_1 s}} = \dfrac{1}{L_1 C_1 s^2 + R_1 C_1 s + 1}$. Output of MATLAB code is shown in Figure 1.257. You can compare different points of graphs in order to ensure that they are the same.

```
Command Window
>> R1=2;L1=50e-6;C1=5e-6;
>> H=tf([1],[L1*C1 R1*C1 1]);
>> impulse(H,0.5e-3)
fx >> |
```

Figure 1.256

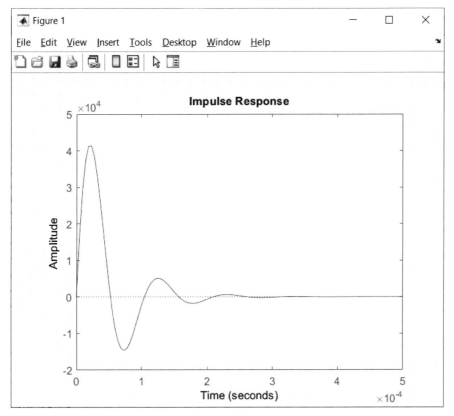

Figure 1.257

1.29 Example 26: Unit Step Response of Circuits

TINA-TI can be used to calculate the unit step response of electric circuits as well. In this example, we want to obtain the unit step response of the RLC circuit in Example 25.

Let's start. Open the schematic of Example 25 (Figure 1.258).

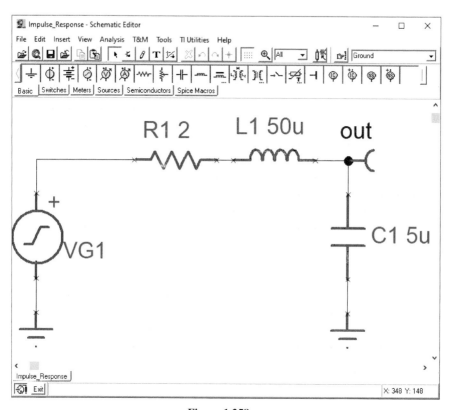

Figure 1.258

Double click the Voltage Generator block VG1 and change its settings to what is shown in Figures 1.259 and 1.260.

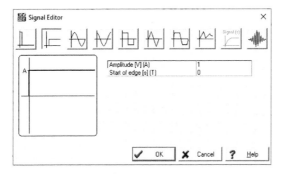

Figure 1.259

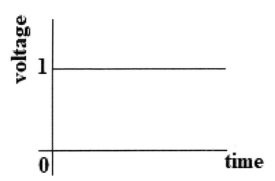

Figure 1.260

The settings shown in Figure 1.260 generate the waveform shown in Figure 1.261.

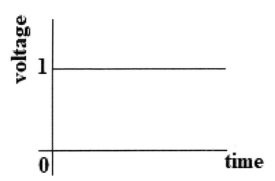

Figure 1.261

Run a transient analysis with the parameters shown in Figure 1.262. The simulation result is shown in Figure 1.263.

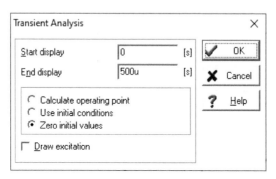

Figure 1.262

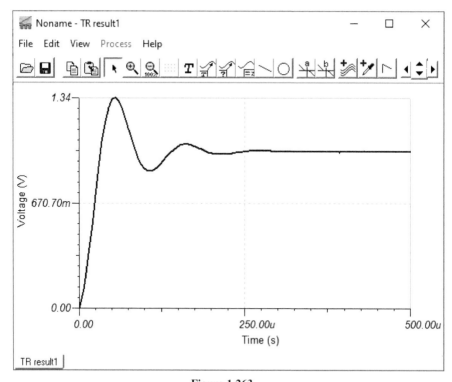

Figure 1.263

Let's check the obtained result. The MATLAB code shown in Figure 1.264 draws the unit step response of the circuit. Note that the transfer function of the RLC circuit in Figure 1.258 is $\frac{V_{C_1}(s)}{V_1(s)} = \frac{\frac{1}{C_1 s}}{R_1 + L_1 s + \frac{1}{C_1 s}} = \frac{1}{L_1 C_1 s^2 + R_1 C_1 s + 1}$. The output of MATLAB code is shown in Figure 1.265. You can compare different points of graphs in order to ensure that they are the same.

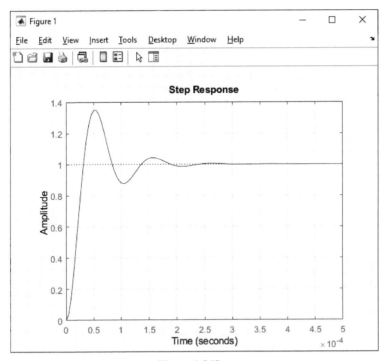

Figure 1.264

Figure 1.265

1.30 Example 27: Frequency Response of Electric Circuits (I)

TINA-TI can be used to calculate the frequency response of electric circuits as well. In this example, we want to obtain the frequency response of the RLC circuit in Example 25. We want to obtain $H(j\omega) = \frac{V_C(j\omega)}{V_{in}(j\omega)}$. $V_C(j\omega)$ and $V_{in}(j\omega)$ show the capacitor voltage and input voltage, respectively.

Let's start. Open the schematic of Example 25 (Figure 1.266). Note that the Signal box (Figure 1.267) has no effect on the frequency response analysis. You can do a frequency response analysis even when the Signal box contains a non-sinusoidal signal.

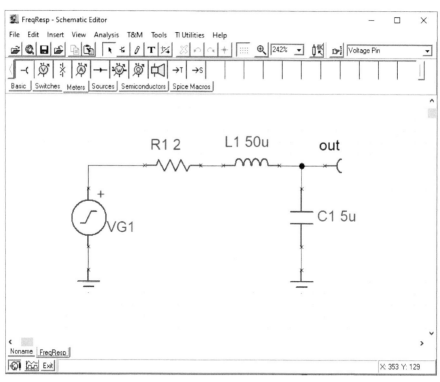

Figure 1.266

VG2 - Voltage Generator		✕
Label	VG2	
Parameters	[Parameters]	
DC Level [V]	0	☐
Signal	Pulse ...	
Internal resistance [Ohm]	0	☐
IO state	Input	
Fault	None	

✓ OK   ✕ Cancel   ? Help

Figure 1.267

Click the Analysis> AC Analysis> AC Transfer Characteristic (Figure 1.268). After clicking, the window shown in Figure 1.269 appears on the screen. You can see the description of each parameter by clicking the Help button (Figure 1.270).

Enter the desired frequency range into the Start frequency and End frequency boxes and click the OK button. The simulation result is shown in Figure 1.271.

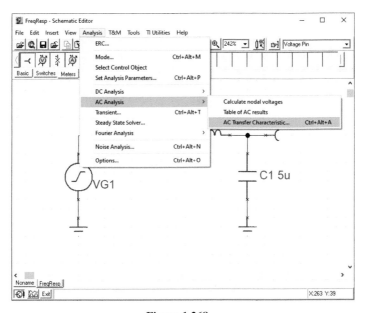

Figure 1.268

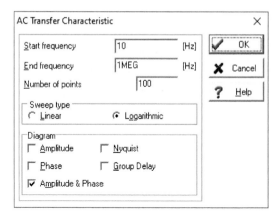

Figure 1.269

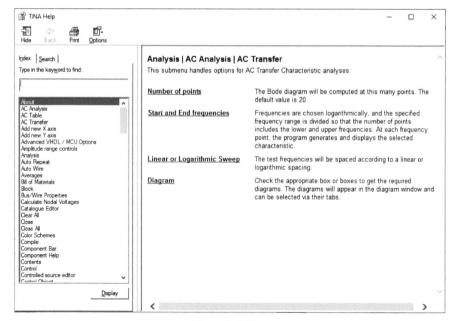

Figure 1.270

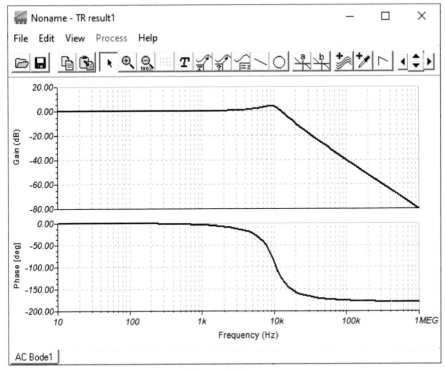

Figure 1.271

Let's check the obtained result. The MATLAB code shown in Figure 1.272 draws the frequency response of the circuit for [10 Hz, 1 MHz] range. The output of MATLAB code is shown in Figure 1.273.

```
Command Window
>> R=2;L=50e-6;C=5e-6;
>> H=tf(1,[L*C R*C 1]);
>> f1=10;
>> f2=1e6;
>> w=logspace(log10(2*pi*f1),log10(2*pi*f2));
>> bode(H),grid on
>> title('VC(s)/Vin(s)')
fx >>
```

Figure 1.272

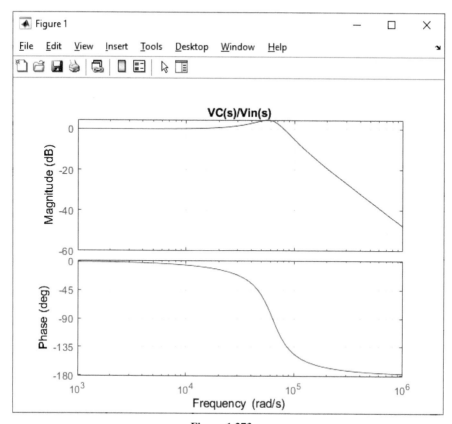

Figure 1.273

Note that the horizontal axis in Figure 1.273 has the unit of $\frac{Rad}{s}$. Let's convert it into Hz. Double click on the white region in Figure 1.273. This opens the window shown in Figure 1.274.

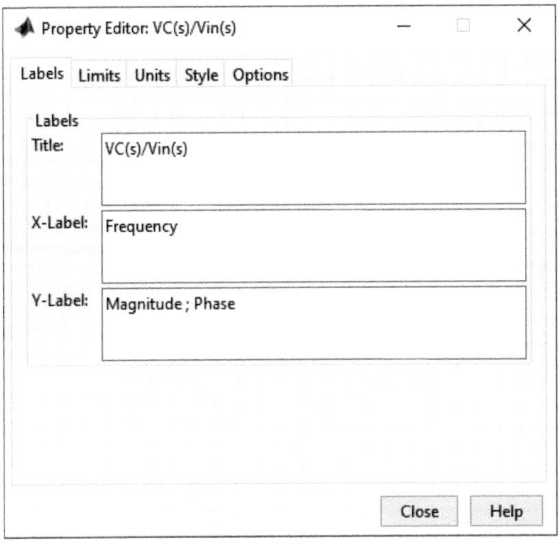

Figure 1.274

Open the Units tab and select Hz for the Frequency box (Figure 1.275). Then click the Close button. Now, the horizontal axis has the unit of Hz (Figure 1.276). Compare different points of the two graphs in Figure 1.271 and Figure 1.276 in order to ensure that both of them are the same.

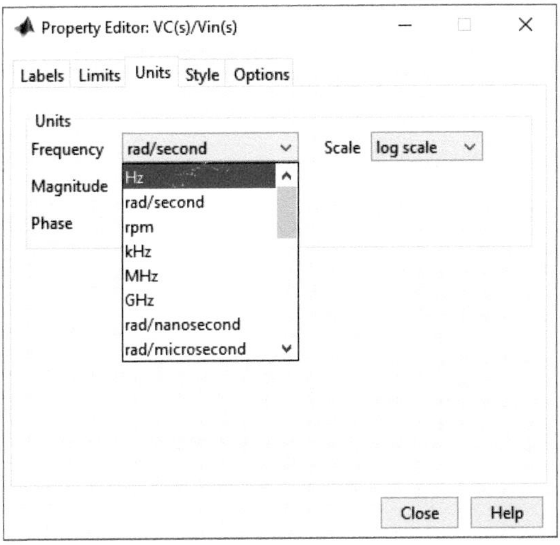

Figure 1.275

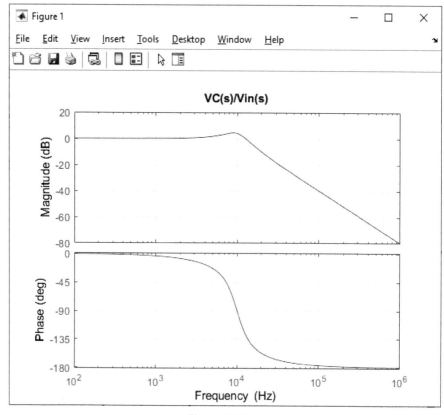

Figure 1.276

1.31 Example 28: Frequency Response of Electric Circuits (II)

In this example we want to obtain the frequency response of $H(j\omega) = \frac{I_{in}(j\omega)}{V_{in}(j\omega)}$ for the RLC circuit in Example 25. $I_{in}(j\omega)$ and $V_{in}(j\omega)$ show the circuit current and input voltage, respectively.

Let's start. Change the schematic of Example 25 to what is shown in Figure 1.277.

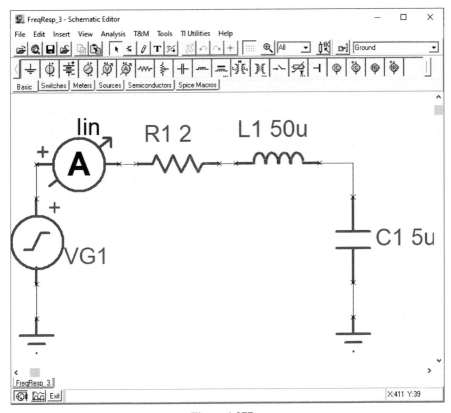

Figure 1.277

Click the Analysis> AC Analysis> AC Transfer Characteristic (Figure 1.278). Do the analysis with the parameters shown in Figure 1.279. The simulation result is shown in Figure 1.280.

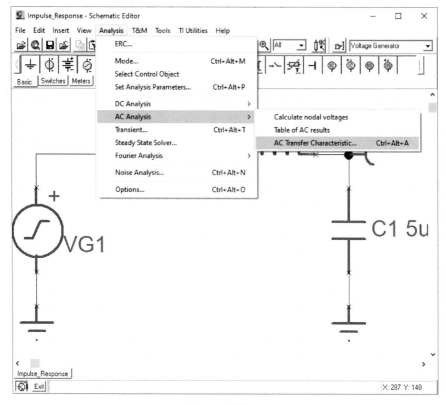

Figure 1.278

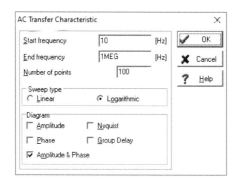

Figure 1.279

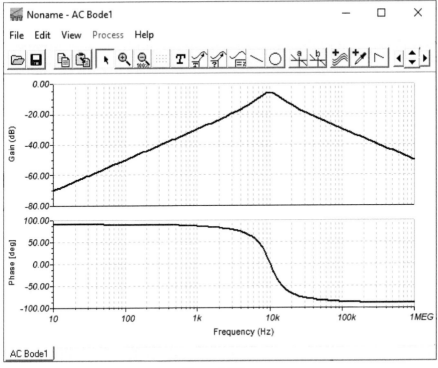

Figure 1.280

Click the File> Export> As Text (Figure 1.281) and save the obtained data with the name 'Current.txt'. This file will be used in the next example.

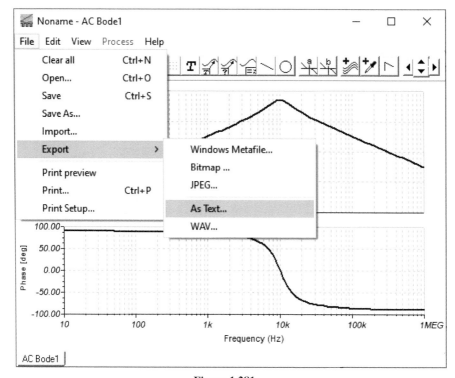

Figure 1.281

Let's check the obtained result. Following MATLAB code draws the frequency response of $H(j\omega) = \frac{I_{in}(j\omega)}{V_{in}(j\omega)}$ for the RLC circuit in Example 25. Note that $H(s) = \frac{I_{in}(s)}{V_{in}(s)} = \frac{Cs}{LCs^2 + RCs + 1}$. You can compare different points of the two graphs in Figure 1.280 and Figure 1.283 in order to ensure that both of them are the same.

```
Command Window
>> R=2;L=50e-6;C=5e-6;
>> H=tf([C 0],[L*C R*C 1]);
>> f1=10;
>> f2=1e6;
>> w=logspace(log10(2*pi*f1),log10(2*pi*f2));
>> bode(H),grid on
fx >>
```

Figure 1.282

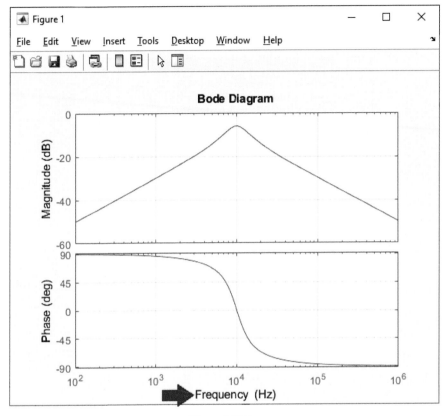

Figure 1.283

1.32 Example 29: Input Impedance of Electric Circuits

In this example, we want to draw the graph of the input impedance of the RLC circuit in Example 25. Let's start. Import the data of Example 28 into MATLAB (Figure 1.284).

Figure 1.284

Following MATLAB code (Figure 1.285) draws the input impedance of the circuit. The output of this code is shown in Figure 1.286.

```
Command Window
>> freq=table2array(Current(:,1));
>> I_mag_dB=table2array(Current(:,2));
>> I_phase=table2array(Current(:,3));
>> subplot(211)
>> semilogx(freq,10.^(-I_mag_dB/20))
>> subplot(212)
>> semilogx(freq,-I_phase)
fx >> |
```

Figure 1.285

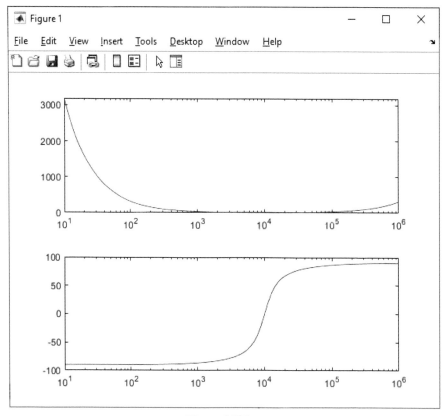

Figure 1.286

You can export the obtained graph as a graphical file by clicking the File>
Save As (Figure 1.287). After clicking the File> Save as the Save as window
appears on the screen. Select the desired output file from Save as a type list
(Figure 1.288).

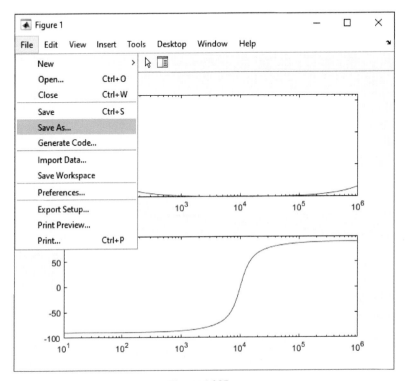

Figure 1.287

Figure 1.288

Let's check the obtained result. According to Figure 1.289, the input impedance at 100 Hz is $318.285e^{-j89.64°} \approx 2 - 318.28j$. The commands shown in Figure 1.290 calculate the input impedance of the RLC circuit at 100 Hz. The obtained result is the same as the result shown in Figure 1.289.

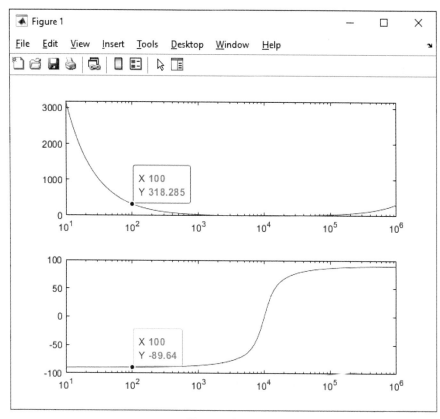

Figure 1.289

```
Command Window                              ⊙
    >> R=2;L=50e-6;C=5e-6;
    >> w=2*pi*100;
    >> Z=R+j*L*w-j/(C*w)

    Z =

        2.0000e+00 - 3.1828e+02i

    >> abs(Z)

    ans =

        318.2848

    >> angle(Z)*180/pi

    ans =

        -89.6400

fx >>
```

Figure 1.290

Example 22 of Chapter 2 introduces another technique to draw the input impedance of a circuit.

1.33 Example 30: Drawing the Input Impedance of Electric Circuits with MATLAB®

In this example we want to use MATLAB to draw the graph of the input impedance of the RLC circuit in Example 25. From basic circuit theory, $Z_{in}(s) = R + Ls + \frac{1}{Cs}$. The MATLAB code shown in Figure 1.291 draws the frequency response of $Z_{in}(s)$ on the [10 Hz, 1 MHz] range. The output of this code is shown in Figure 1.292.

```
Command Window                                        ⊙
    >> R=2;L=50e-6;C=5e-6;
    >> s=tf('s');
    >> H=R+L*s+1/C/s;
    >> f1=10;
    >> f2=1e6;
    >> w=logspace(log10(2*pi*f1),log10(2*pi*f2));
    >> bode(H,w)
fx >>
```

Figure 1.291

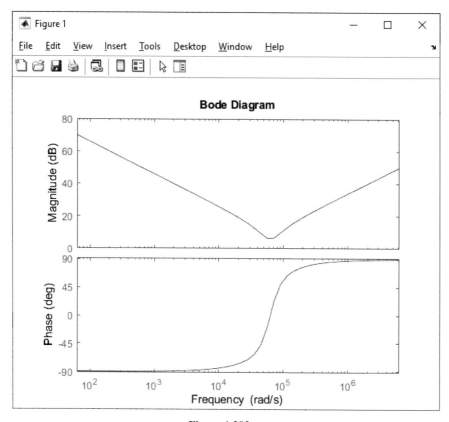

Figure 1.292

Note that in Figure 1.292, the vertical axis has the unit of dB (i.e., $20\log(|Z_{in}(j\omega)|)$). dB is not the common unit for input impedance graphs. So, it is better to change it into normal a gain (i.e., $|Z_{in}(j\omega)|$). The horizontal axis has the unit of $\frac{Rad}{s}$ which needs to be converted into Hz. Let's do these changes.

Double click on the white region of Figure 1.292. This opens the Property Editor window (Figure 1.293).

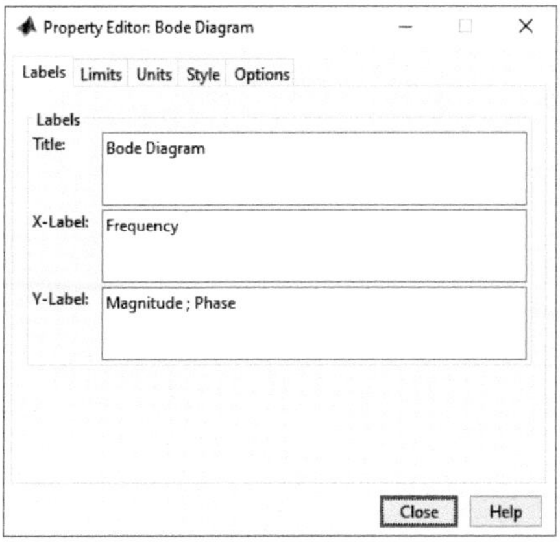

Figure 1.293

Open the Units tab and do the settings similar to Figure 1.294. The diagram changes to what is shown in Figure 1.295. Now, the vertical axis is not in dB, it is in Ohm.

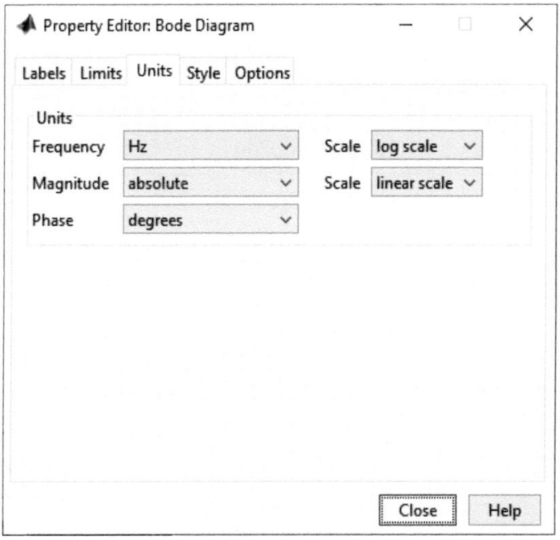

Figure 1.294

Note that the graph in Figure 1.295 is the same as the graph in Figure 1.286. You can ensure by comparing different points together.

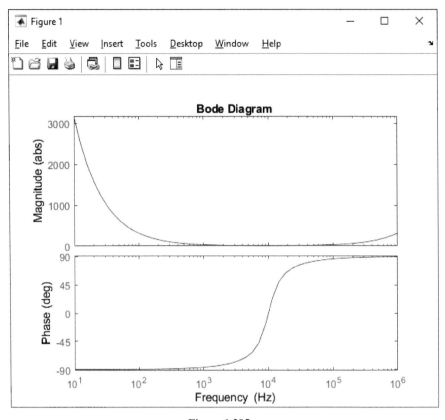

Figure 1.295

1.34 Example 31: Phasor Analysis

Phasor analysis can be done quite easily in TINA-TI. Let's study an example. Draw the schematic shown in Figure 1.296. Settings of Voltage Generator block VG1 are shown in Figure 1.297. According to Figure 1.297, the input voltage is $169.7\sin\left(\omega t + 0°\right) = 169.7\cos\left(\omega t - \frac{\pi}{2}\right)$. So, phasor of input voltage is $169.7\angle - \frac{\pi}{2} = 169.7^{-j\frac{\pi}{2}}$.

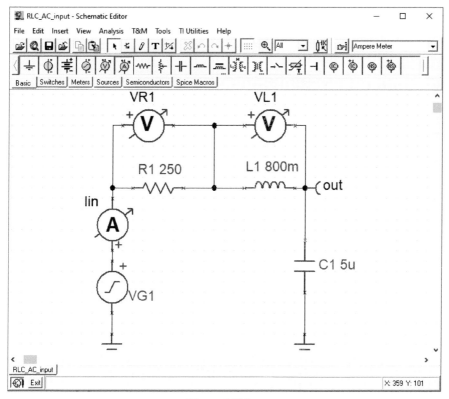

Figure 1.296

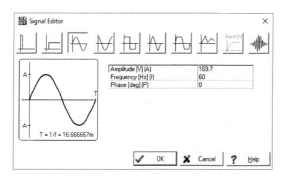

Figure 1.297

Click the Analysis> AC Analysis> Calculate nodal voltages (Figure 1.298). The simulation result is shown in Figure 1.299.

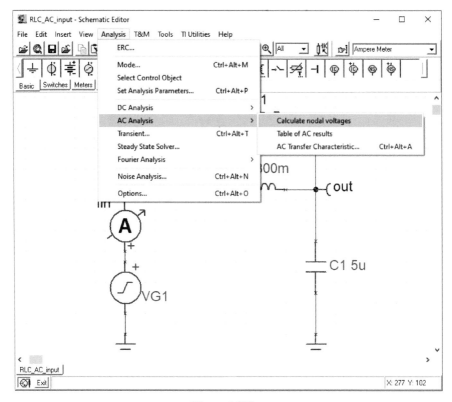

Figure 1.298

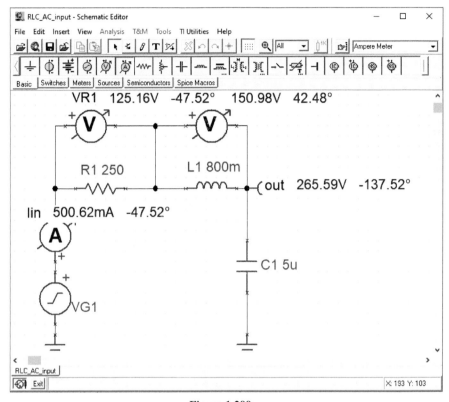

Figure 1.299

According to Figure 1.299, the circuit current is $0.50062\cos$ $(\omega t - 47.52°)$ *A.* Let's check this result. MATLAB calculations in Figure 1.300 show that the TINA-TI result is correct.

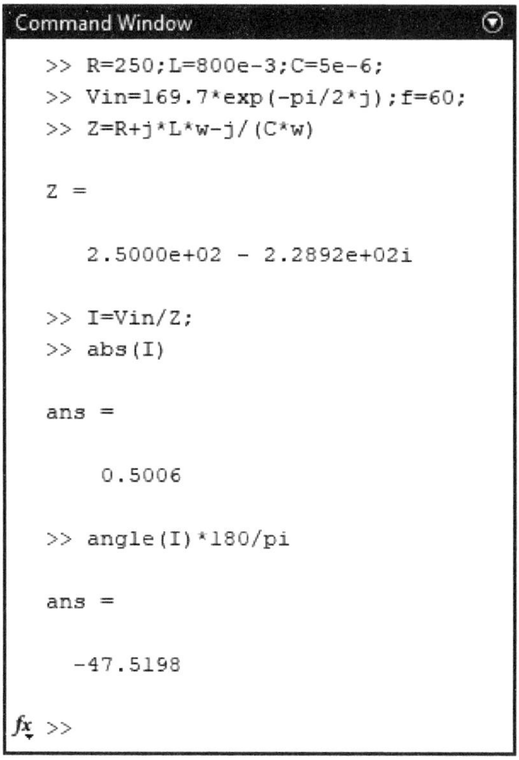

```
Command Window                              ⊙
  >> R=250;L=800e-3;C=5e-6;
  >> Vin=169.7*exp(-pi/2*j);f=60;
  >> Z=R+j*L*w-j/(C*w)

  Z =

      2.5000e+02 - 2.2892e+02i

  >> I=Vin/Z;
  >> abs(I)

  ans =

        0.5006

  >> angle(I)*180/pi

  ans =

    -47.5198

fx >>
```

Figure 1.300

According to Figure 1.299, a voltage of the capacitor is $265.59\cos$ $(\omega t - 137.52°)$ V. Let's check this result. MATLAB calculations in Figure 1.301 show that the TINA-TI result is correct.

```
Command Window                                          ⊙
    >> R=250;L=800e-3;C=5e-6;
    >> Vin=169.7*exp(-pi/2*j);f=60;
    >> Z=R+j*L*w-j/(C*w);
    >> Zc=-j/(C*w);
    >> abs(Zc/Z*Vin)

  ans =

    265.5886

    >> angle(Zc/Z*Vin)*180/pi

  ans =

    -137.5198

fx >> |
```

Figure 1.301

1.35 Example 32: Parameter Sweep Analysis

Parameter sweep analysis permits you to change the value of a parameter and see its effect on the circuit behavior. For instance, assume that we want to change the value of capacitor C1 (Figure 1.302) from 4 μF up to 20 μF with 4 μF steps and see its effect on the frequency response $H(j\omega) = \frac{V_{C1}(j\omega)}{V_{in}(j\omega)}$ of the circuit.

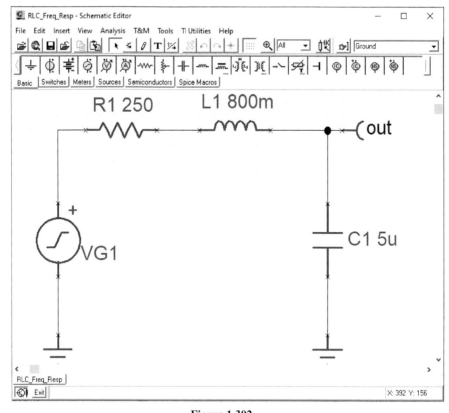

Figure 1.302

Let's start. Click the Analysis> Select Control Object (Figure 1.303). Then click on the capacitor C1. This opens the window shown in Figure 1.304.

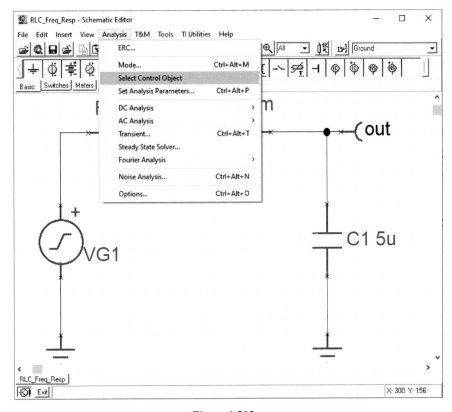

Figure 1.303

Figure 1.304

Click the Select button. This opens the Control object selection window. In this example, we want to change the value of capacitor C1 from 4 μF up to 20 μF with 4 μF steps. In other words, we want to analyze the circuit for C1= 4 μF, 8 μF, 12 μF, 16 μF and 20 μF. So, Start value= 4 μF, End value= 20 μF and Number of cases= 5 (Figure 1.305).

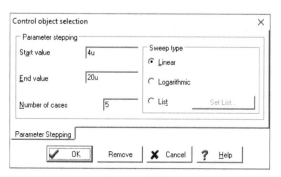

Figure 1.305

After clicking the OK button in Figure 1.305, a star appears behind capacitor C1 (Figure 1.306).

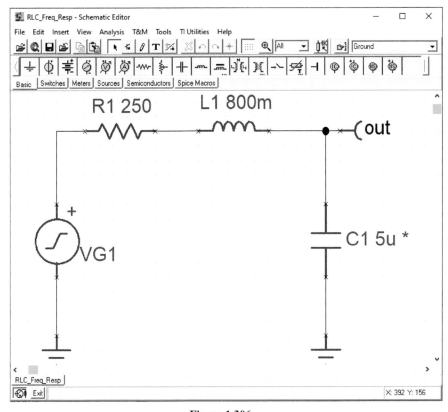

Figure 1.306

Click the Analysis> AC Analysis> AC Transfer Characteristic
(Figure 1.307). Run the frequency response simulation with the parameters
shown in Figure 1.308. These settings draw the frequency response for the
[10 Hz, 200 Hz] range. The simulation result is shown in Figure 1.309.

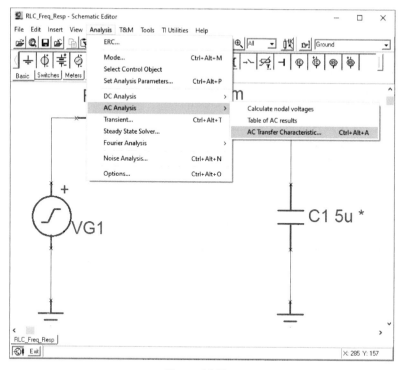

Figure 1.307

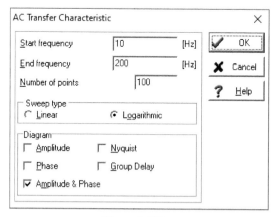

Figure 1.308

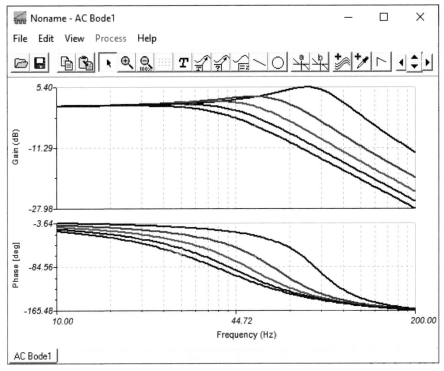

Figure 1.309

You can easily see which graph belongs to which value of the capacitor. Just right-click on the graph and click the Modified components (Figure 1.310). This opens the Components Parameters window and shows the value of the capacitor for that curve (Figure 1.311). The frequency response becomes more flat as the value of the capacitor increases.

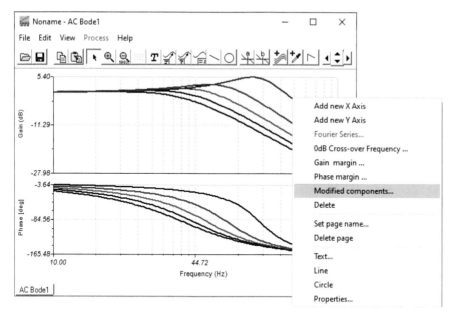

Figure 1.310

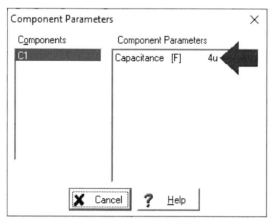

Figure 1.311

The addition of legend to the graph is another way to see which curve belongs to which value of a parameter (Figure 1.312).

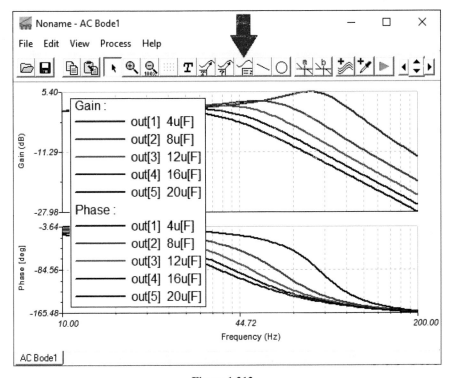

Figure 1.312

Let's study the effect of capacitor value on the transient response of the circuit. Close the Figure 1.312 and click the Analysis> Transient (Figure 1.313). Run the transient analysis with the parameters shown in Figure 1.314. The simulation result is shown in Figure 1.315. When the capacitor C1 value increases, the amplitude of output decreases.

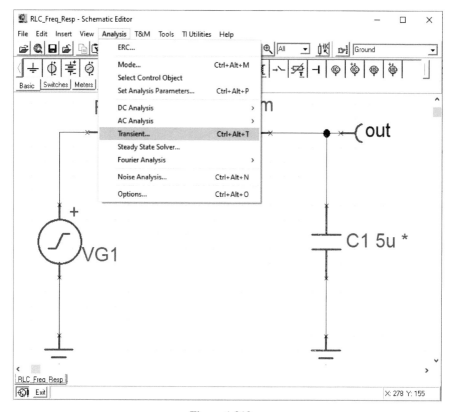

Figure 1.313

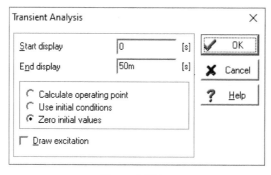

Figure 1.314

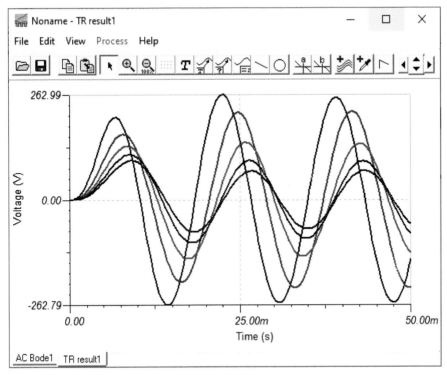

Figure 1.315

1.36 Exercises

1.a) Find the Thevenin equivalent circuit with respect to the terminals 'a' and 'b' for the circuit shown in Figure 1.316 and Figure 1.317.
   b) Use TINA-TI to check the result of part (a).

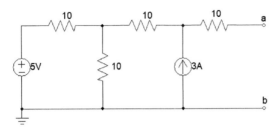

Figure 1.316

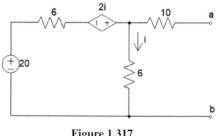

Figure 1.317

2. Simulate the circuit shown in Figure 1.318. Initial conditions are shown on the Figure.

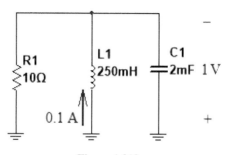

Figure 1.318

3. In the circuit shown in Figure 1.319, $V_1 = 10 + 25\sin(2\pi \times 60t)$. Initial conditions are $V_{C,0} = 10\ V$ and $i_{L,0} = 0\ A$. Use TINA-TI to observe the circuit current.

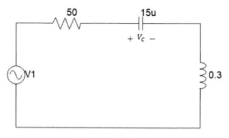

Figure 1.319

4.a) Calculate the current i in the circuit of Figure 1.320.
  b) Use TINA-TI to check the result of part (a).

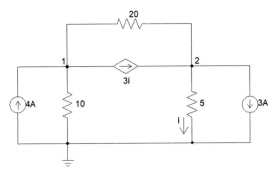

Figure 1.320

5. Set up a TINA-TI simulation to measure the RMS of a triangular wave. Use MATLAB or hand calculation to verify the obtained result.
6. In Figure 1.321, 30< Rload <40. Use TINA-TI to find the value of the load resistor R load which consumes the maximum power (**Hint:** Use parameter sweep analysis).

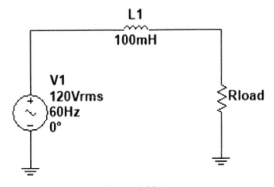

Figure 1.321

7. Use TINA-TI to observe the unit impulse response and unit step response of the circuit shown in Figure 1.322 (output is the capacitor voltage).

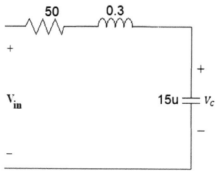

Figure 1.322

8. Use TINA-TI to observe the frequency response ($\frac{i(j\omega)}{V_{in}(j\omega)}$) of the circuit
   shown in Figure 1.323 (output is the inductor current).

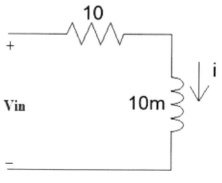

Figure 1.323

References

[1] Hayt, W., Kemmerly, J., and Durbin, S.: Engineering circuit analysis, 9th
    edition, McGraw-Hill (2021)
[2] Nilsson, J., and Riedel, S.: Electric circuits, 11th edition, Pearson (2018)
[3] Thomas, R. E., Rosa, A. J., and Toussaint G. J.: The analysis and design
    of linear circuits, 9th edition, John Wiley and Sons (2020)
[4] Alexander, C., and Sadiku, M. N. O.: Fundamentals of electric circuits,
    6th edition, McGraw-Hill (2016)

2

Simulation of Electronic Circuits with TINA-TI®

2.1 Introduction

In this chapter, you will learn how to analyze electronic circuits in TINA-TI. The theory behind the studied circuits can be found in any standard electronic/microelectronic textbook [1–3]. Similar to the previous chapter, doing some hand calculations for the given circuits and comparing the hand analysis results with TINA-TI results are recommended.

2.2 Example 1: Half Wave Rectifier

In this example, we want to simulate a half-wave rectifier with a purely resistive load. Draw the schematic shown in Figure 2.1. The diode block can be found in the Semiconductor tab (Figure 2.2).

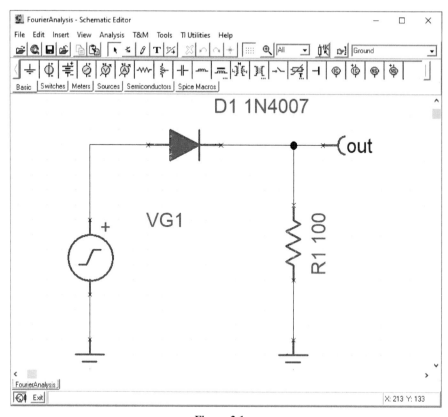

Figure 2.1

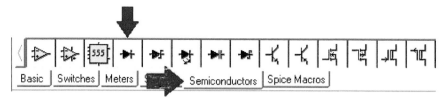

Figure 2.2

Settings of the input voltage source is shown in Figure 2.3.

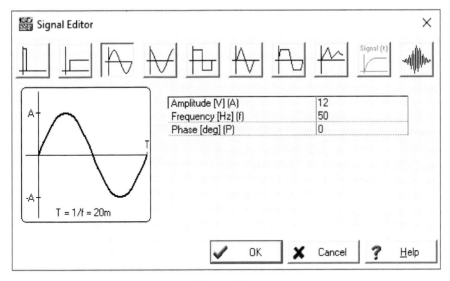

Figure 2.3

Double click on the diode you added to the schematic. This opens the window shown in Figure 2.4. Click the three-dot behind the Type box. This opens the Catalog Editor window (Figure 2.5). Now you can select the diode that you want. In this example we use 1N4007.

D1 - Diode			
Label	D1		
Parameters	(Parameters)		
Type	1N4007	...	✔
Temperature	Relative		
Temperature [°C]	0		☐
Area factor	1		☐
Device initially OFF(DC)	No		
Initial voltage (TR)	Not Used		☐
Fault	None		

Figure 2.4

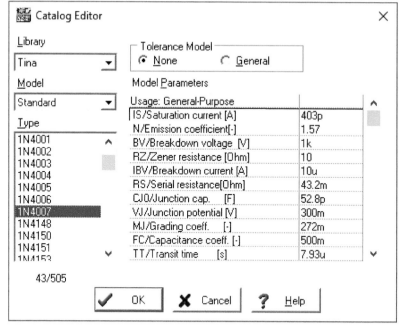

Figure 2.5

Run a transient analysis with the parameters shown in Figure 2.6. The simulation result is shown in Figure 2.7.

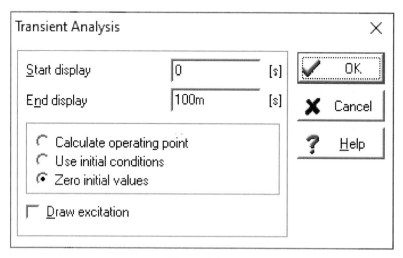

Figure 2.6

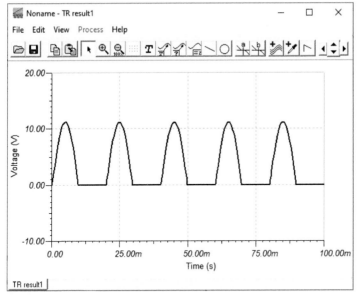

Figure 2.7

Let's measure the peak of the load voltage. According to Figure 2.8, the peak of the load voltage is 11.19 V. So, a voltage drop of the diode is $12 - 11.19 = 0.81$ V.

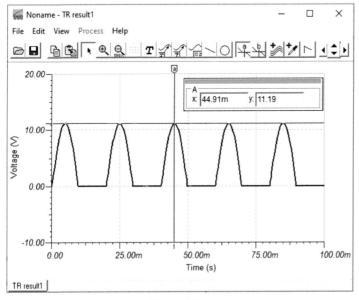

Figure 2.8

Let's measure the frequency of output. According to Figure 2.9, one cycle takes $19.93\ ms \approx 20\ ms$. Therefore, frequency is $\frac{1}{20\ ms} = 50\ Hz$.

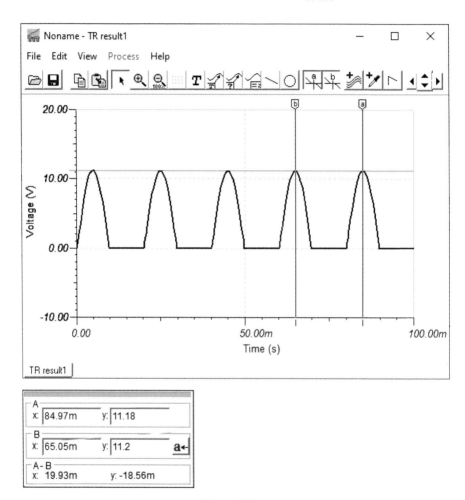

Figure 2.9

2.3 Example 2: Measurement of Average and RMS Values of Waveforms

In the previous example, we simulated a half-wave rectifier a nd we observed its output waveform. In this example, we want to measure the average and

RMS values of the waveform shown in Figure 2.10. Remember that the average value is defined as $\frac{1}{T}\int_{t_0}^{t_0+T} f(\tau)d\tau$, an absolute average value is defined as $\frac{1}{T}\int_{t_0}^{t_0+T} |f(\tau)| d\tau$ and RMS value is defined as $\sqrt{\frac{1}{T}\int_{t_0}^{t_0+T} f(\tau)^2 d\tau}$. T shows the period of the waveform ($f(t+T) = f(t)$).

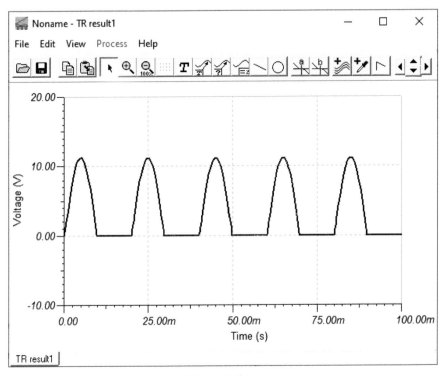

Figure 2.10

Click on the waveform. After clicking on the waveform, its color changes into red (Figure 2.11).

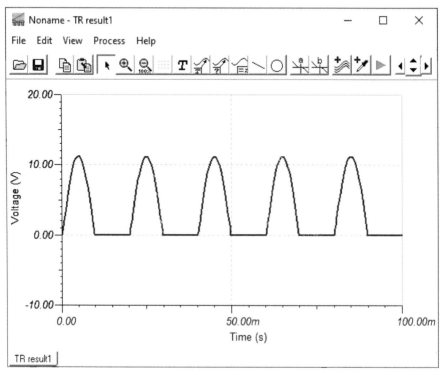

Figure 2.11

Click the Process> Averages (Figure 2.12). After clicking, the window shown in Figure 2.13 appears on the screen. Let's call the waveform shown in Figure 2.12, $f(t)$. According to Figure 2.13, $\frac{1}{100ms-0s}\int_{0s}^{100ms}f(t)dt = 3.44\ V$, $\frac{1}{100ms-0s}\int_{0s}^{100ms}|f(t)|\,dt = 3.44\ V$ and $\sqrt{\frac{1}{100ms-0s}\int_{0s}^{100ms}f(t)^2dt} = 5.51\ V$.

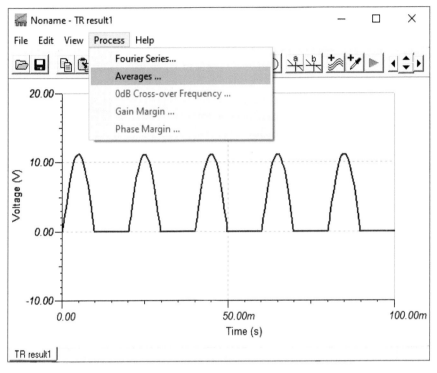

Figure 2.12

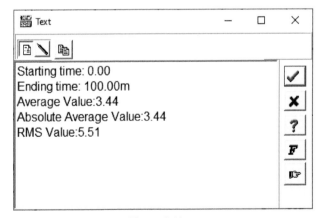

Figure 2.13

The above-mentioned method (i.e., clicking on the waveforms and clicking the Process> Averages) enters the whole waveform into the calculation. However, accurate measurement of average and RMS values must be done with the steady-state portion of the waveform only. In this example, the load is resistive. Therefore, we have no transient region and the above-mentioned method works perfectly. However, in problems that have transient regions, it is recommended to use the magnifier icon and select one cycle from the steady-state region of the waveform. Let's use this method to measure the average and RMS of the waveform shown in Figure 2.12. In Figure 2.14, one cycle of the output is selected.

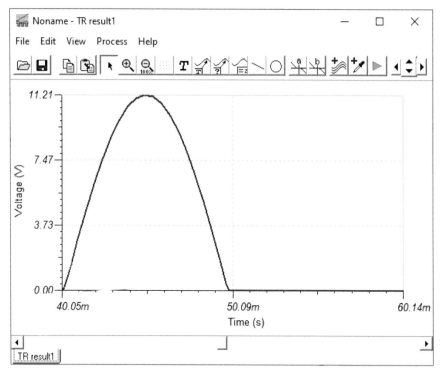

Figure 2.14

Click the Process> Averages (Figure 2.15). After clicking, the window shown in Figure 2.16 appears on the screen. According to Figure 2.16, $\frac{1}{59.85ms-40.15ms}\int_{40.15ms}^{59.85ms} f(t)dt = 3.49 \ V$, $\frac{1}{59.85ms-40.15ms}\int_{40.15ms}^{59.85ms} |f(t)| dt = 3.49 \ V$ and $\sqrt{\frac{1}{59.85ms-40.15ms}\int_{40.15ms}^{59.85ms} f(t)^2 dt} = 5.55 \ V$. Obtained results are quite close to the results shown in Figure 2.13 since we have no transient region.

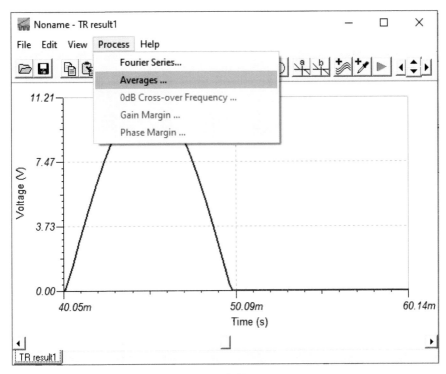

Figure 2.15

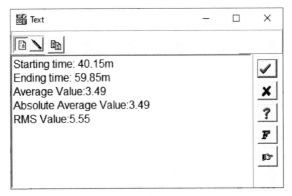

Figure 2.16

2.4 Example 3: Harmonic Content of Waveforms

TINA-TI can show the harmonic content of the waveforms easily. As an example let's see the harmonic content of Example 2. Right-click on the waveform and click the Fourier Series (Figure 2.17). This opens the Fourier Series window (Figure 2.18). The description of parameters in this window can be seen by clicking the Help button (Figure 2.19).

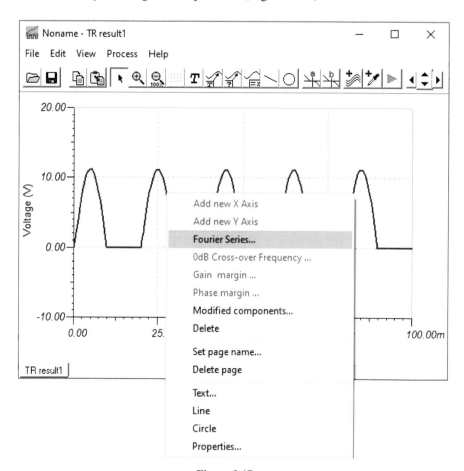

Figure 2.17

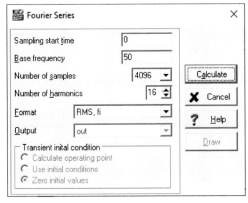

Figure 2.18

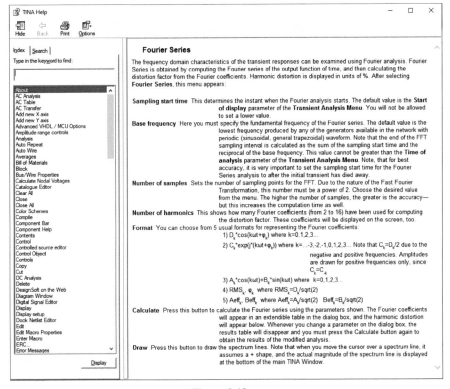

The frequency domain characteristics of the transient responses can be examined using Fourier analysis. Fourier Series is obtained by computing the Fourier series of the output function of time, and then calculating the distortion factor from the Fourier coefficients. Harmonic distortion is displayed in units of %. After selecting **Fourier Series**, this menu appears:

Sampling start time This determines the instant when the Fourier analysis starts. The default value is the **Start of display** parameter of the **Transient Analysis Menu**. You will not be allowed to set a lower value.

Base frequency Here you must specify the fundamental frequency of the Fourier series. The default value is the lowest frequency produced by any of the generators available in the network with periodic (sinusoidal, general trapezoidal) waveform. Note that the end of the FFT sampling interval is calculated as the sum of the sampling start time and the reciprocal of the base frequency. This value cannot be greater than the **Time of analysis** parameter of the **Transient Analysis Menu**. Note, that for best accuracy, it is very important to set the sampling start time for the Fourier Series analysis to after the initial transient has died away.

Number of samples Sets the number of sampling points for the FFT. Due to the nature of the Fast Fourier Transformation, this number must be a power of 2. Choose the desired value from the menu. The higher the number of samples, the greater is the accuracy— but this increases the computation time as well.

Number of harmonics This shows how many Fourier coefficients (from 2 to 16) have been used for computing the distortion factor. These coefficients will be displayed on the screen, too.

Format You can choose from 5 usual formats for representing the Fourier coefficients:

1) $D_k*\cos(k\omega t+\varphi_k)$ where k=0,1,2,3...

2) $C_k*\exp(j*(k\omega t+\varphi_k))$ where k=...-3,-2,-1,0,1,2,3... Note that $C_k=D_k/2$ due to the negative and positive frequencies. Amplitudes are drawn for positive frequencies only, since
$$C_k=C_{-k}$$

3) $A_k*\cos(k\omega t)+B_k*\sin(k\omega t)$ where k=0,1,2,3...

4) RMS_k, φ_k where $RMS_k=D_k/sqrt(2)$

5) $Aeff_k$, $Beff_k$ where $Aeff_k=A_k/sqrt(2)$ $Beff_k=B_k/sqrt(2)$

Calculate Press this button to calculate the Fourier series using the parameters shown. The Fourier coefficients will appear in an extendible table in the dialog box, and the harmonic distortion will appear below. Whenever you change a parameter on the dialog box, the results table will disappear and you must press the Calculate button again to obtain the results of the modified analysis.

Draw Press this button to draw the spectrum lines. Note that when you move the cursor over a spectrum line, it assumes a + shape, and the actual magnitude of the spectrum line is displayed at the bottom of the main TINA Window.

Figure 2.19

The base frequency box shows the (fundamental) frequency of the wave-form. According to Figure 2.9, the frequency of the waveform is 50 Hz. A number of harmonics boxes determines the number of harmonics that must be calculated by TINA-TI. Use the Format box (Figure 2.20) to determine the desired format for the Fourier series. In this example, we want to see the amplitude of harmonics. So, 'D*cos(kwt+fi)' is a good option for us. If you want to see the RMS value of harmonics, select the 'RMS, fi'.

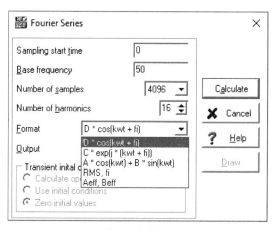

Figure 2.20

Click the Calculate button. The result shown in Figure 2.21 appears on the screen. Note that TINA-TI calculates the Total Harmonic Distortion (THD) as well.

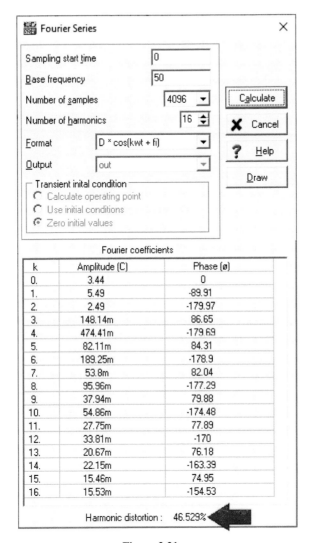

Figure 2.21

If you click the Draw button in Figure 2.22, the graph shown in Figure 2.22 appears on the screen. This graph shows the amplitude and phase of each harmonic. According to Figure 2.22, DC, fundamental and second harmonic amplitude is considerable. Other harmonics can be ignored.

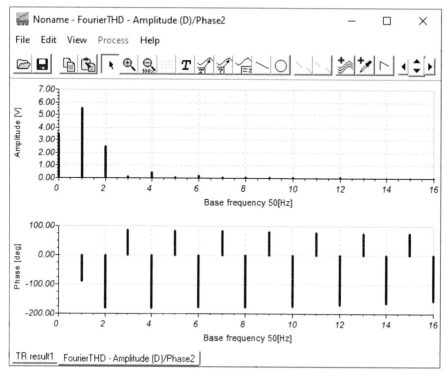

Figure 2.22

Let's check the obtained results. Fourier series of half-wave rectified waveform Fourier series of half wave rectified waveform is shown in Figure 2.23.

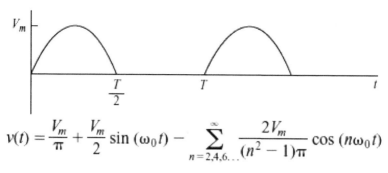

$$v(t) = \frac{V_m}{\pi} + \frac{V_m}{2} \sin(\omega_0 t) - \sum_{n=2,4,6\ldots}^{\infty} \frac{2V_m}{(n^2 - 1)\pi} \cos(n\omega_0 t)$$

Figure 2.23

Calculations shown in Figure 2.24 calculate the amplitude of DC, funda-mental, second, third and fourth harmonics. Obtained results are quite close to TINA-TI results shown in Figure 2.21. Note that the formula in Figure 2.23 is extracted for a half-wave rectifier with an ideal diode (i.e., a diode with zero forward bias voltage drops). The diode in this example is not ideal. In other words, what is shown in Figure 2.23 is only an approximation for the load voltage shown in Figure 2.10. So, a little bit of difference between theory (Figure 2.24) and TINA-TI result (Figure 2.21) is expected.

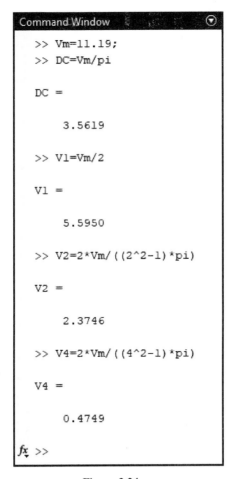

Figure 2.24

2.5 Example 4: Fourier Analysis

In the previous example, we ran a transient analysis and we saw how to obtain the harmonic content of the output waveform. You can obtain the harmonic content without running the transient analysis as well.

Let's start. Open the schematic of Example 3.

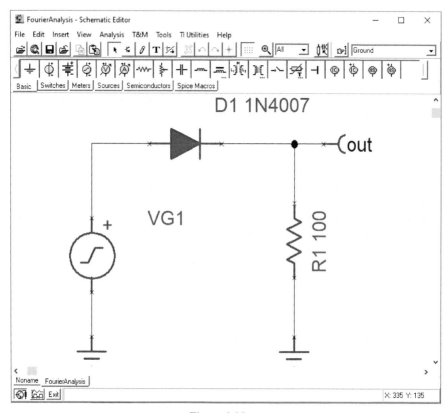

Figure 2.25

Click the Analysis > Fourier Analysis> Fourier Series (Figure 2.26). This opens the Fourier Series window (Figure 2.27). Use the Output box to determine the signal that must be analysed. Click the Calculate button to start the analysis. The simulation result is shown in Figure 2.28. Obtained result is the same as previous example.

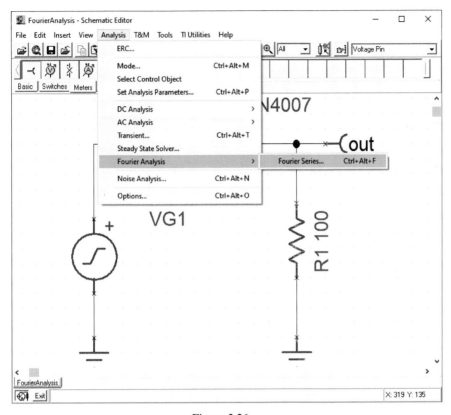

Figure 2.26

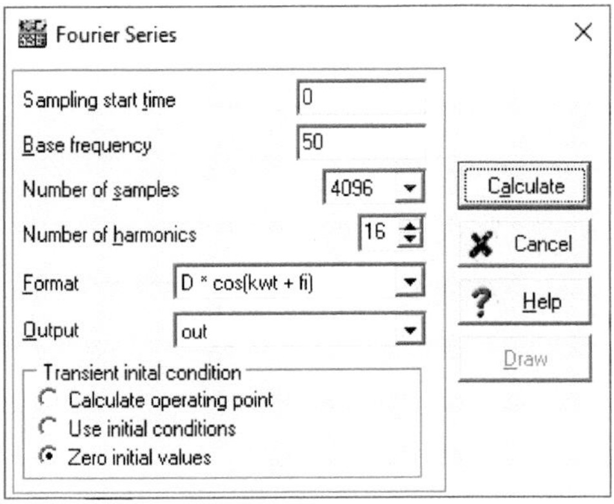

Figure 2.27

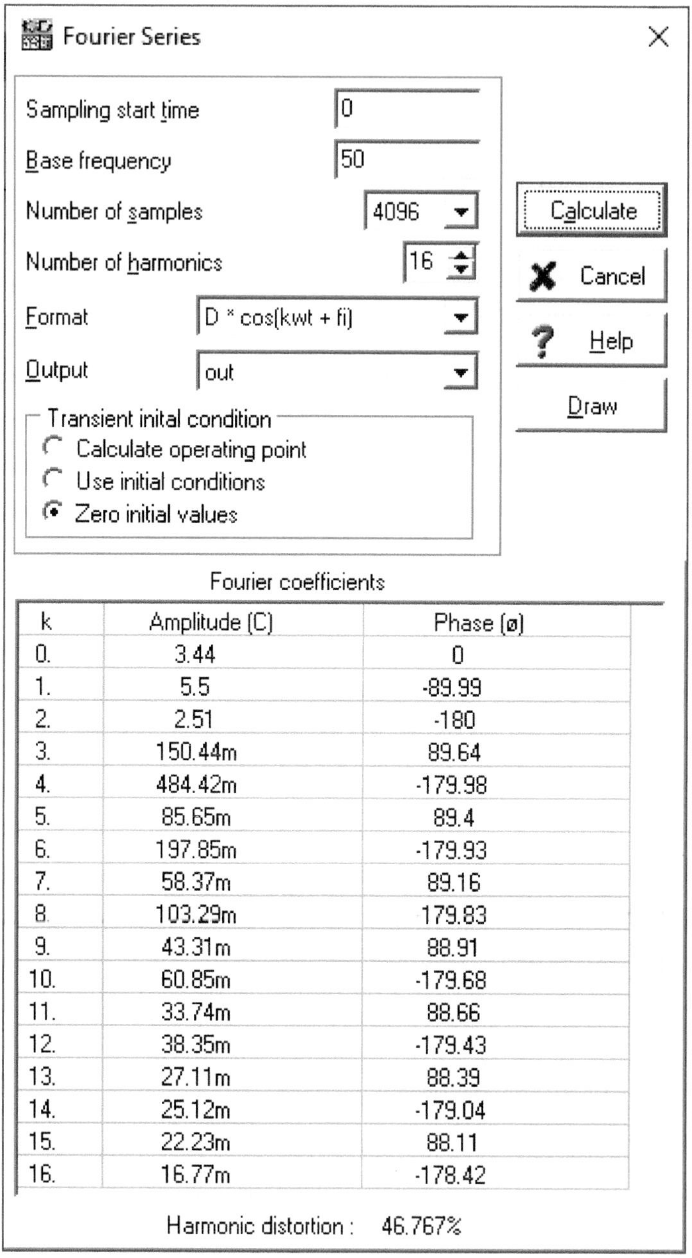

Fourier Series ✕

Sampling start time	0
Base frequency	50
Number of samples	4096 ▼
Number of harmonics	16 ⬍
Format	D * cos(kwt + fi) ▼
Output	out ▼

Calculate

✘ Cancel

? Help

Draw

Transient inital condition
- ○ Calculate operating point
- ○ Use initial conditions
- ⦿ Zero initial values

Fourier coefficients

k	Amplitude (C)	Phase (ø)
0.	3.44	0
1.	5.5	-89.99
2.	2.51	-180
3.	150.44m	89.64
4.	484.42m	-179.98
5.	85.65m	89.4
6.	197.85m	-179.93
7.	58.37m	89.16
8.	103.29m	179.83
9.	43.31m	88.91
10.	60.85m	-179.68
11.	33.74m	88.66
12.	38.35m	-179.43
13.	27.11m	88.39
14.	25.12m	-179.04
15.	22.23m	88.11
16.	16.77m	-178.42

Harmonic distortion : 46.767%

Figure 2.28

2.6 Example 5: Converting a Waveform into Sound

The Play Sound icon (Figure 2.29) can be used to convert a waveform into sound. Click on the waveform and click the Play Sound icon in order to hear the sound associated with the waveform.

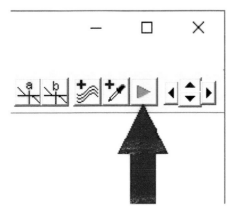

Figure 2.29

2.7 Example 6: DC Transfer Characteristics (I)

DC transfer characteristics permit you to change a parameter and see its effect on the output(s) that you want.

Let's study an example. Consider the schematic shown in Figure 2.30. We want to sweep the value of the input voltage source VS1 from 0 V to 5 V with 50 mV steps and see its effects on the voltage of node 'out'.

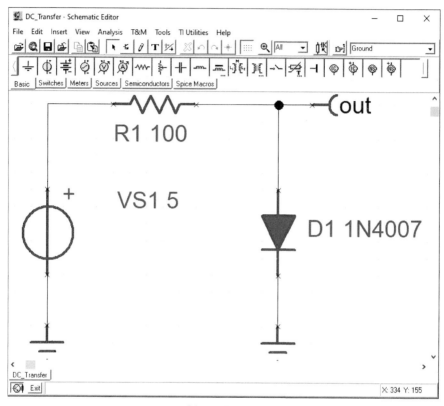

Figure 2.30

Click the Analysis > DC Analysis> DC Transfer Characteristic
(Figure 2.31). This opens the DC Transfer Characteristic window
(Figure 2.32). Settings in figure 2.32 change the value of VS1 from Start value
(=0 V) to End value (=5 V) with steps of $\frac{\text{End value}-\text{Start value}}{\text{Number of points}}$ (=50 mV). Click
the OK button in Figure 2.32 to run the simulation.

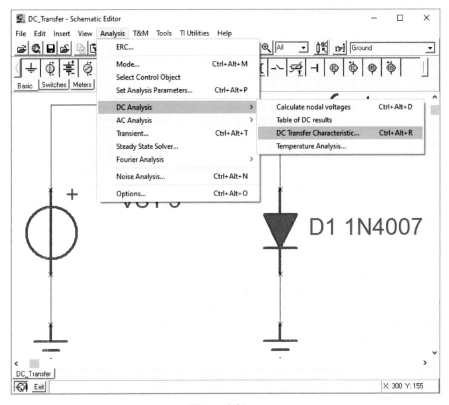

Figure 2.31

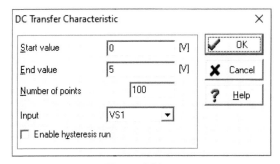

Figure 2.32

Simulation results are shown in Figure 2.33. The vertical axis shows the value of an independent variable (VS1) and the horizontal axis shows the value of the dependent variable (voltage of node 'out').

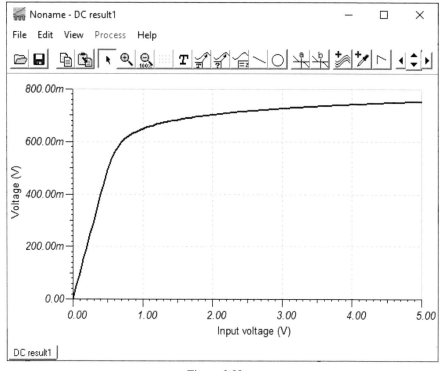

Figure 2.33

2.8 Example 7: DC Transfer Characteristics (II)

In this example, we want to replace the diode of Example 6 with a Zener diode (Figure 2.34) and redraw the DC transfer characteristic of the circuit.

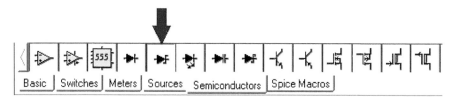

Figure 2.34

Open the schematic of Example 6 and change it to what shown in Figure 2.35.

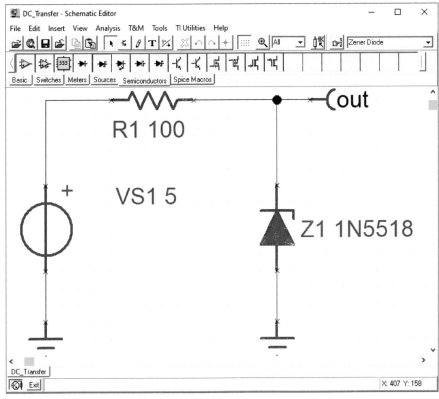

Figure 2.35

Click the Analysis > DC Analysis > DC Transfer Characteristic (Figure 2.36). This opens the DC Transfer Characteristic window (Figure 2.37). Do the settings similar to figure 2.37 (these settings sweep the value of VS1 from 0 V to 5 V with 50 mV steps) and click the OK button. The simulation result is shown in Figure 2.38. The slope of the curve is unity at low voltages. The Zener diode does not conduct in this region. When the input voltage reaches the diode's reverse breakdown voltage, the Zener diode conducts and the slope of the curve starts to decrease.

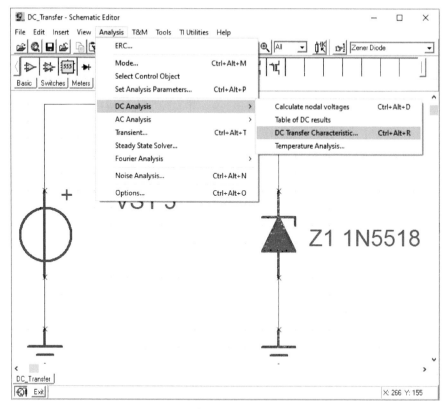

Figure 2.36

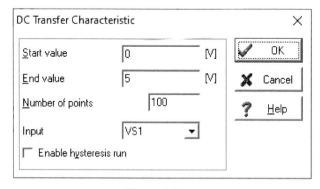

Figure 2.37

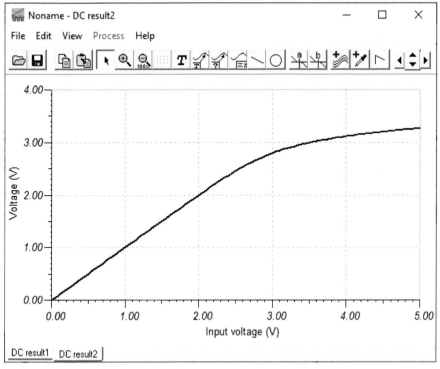

Figure 2.38

2.9 Example 8: DC Transfer Characteristics (III)

In this example, we want to draw the DC transfer characteristic of the circuit shown in Figure 2.39. The input voltage source VS1 changes from −10 V to +10 V with 200 mV steps.

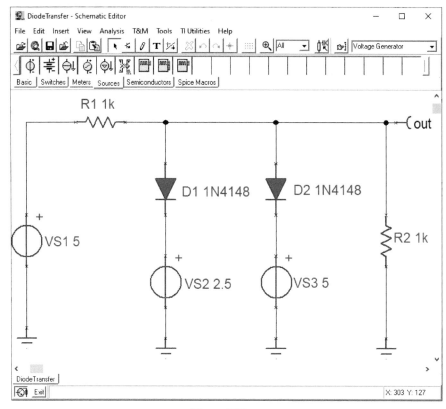

Figure 2.39

Click the Analysis> DC Analysis> DC Transfer Characteristic (Figure 2.40). This opens the DC Transfer Characteristic window. Do the settings similar to Figure 2.41 and click the OK button. The simulation result is shown in Figure 2.42.

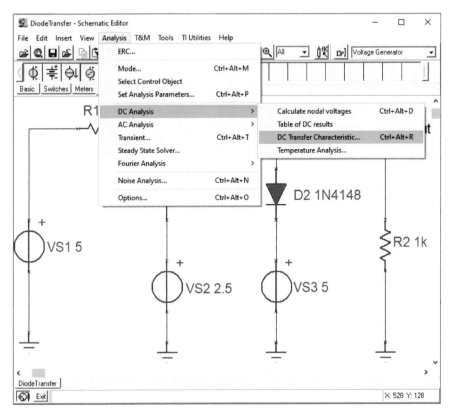

Figure 2.40

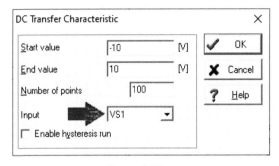

Figure 2.41

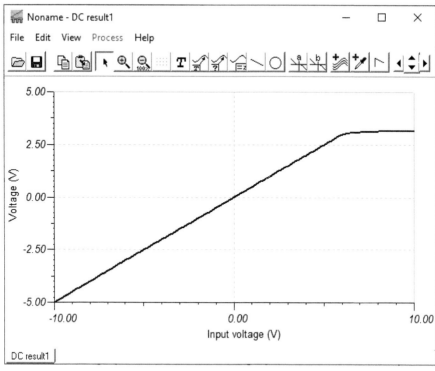

Figure 2.42

2.10 Example 9: Temperature Analysis

TINA-TI does the simulations at 27 °C by default. The default temperature value can be changed. You can do this by clicking the Analysis> Set Analysis Parameters (Figure 2.43). This opens the Analysis Parameters window (Figure 2.44). Enter the desired new value to the Temperature of environment (C) box (Figure 2.44) and click the OK button.

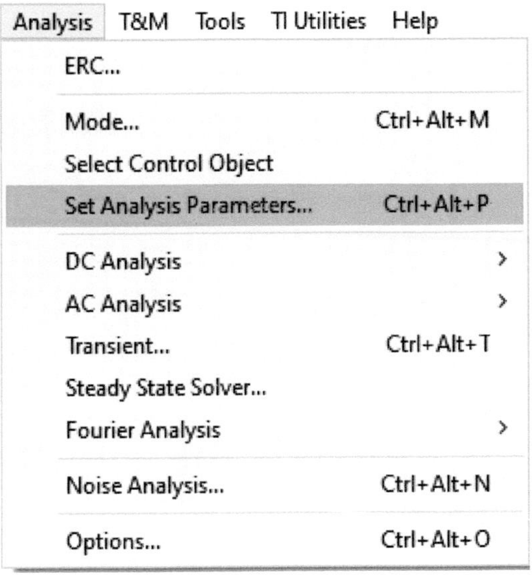

Figure 2.43

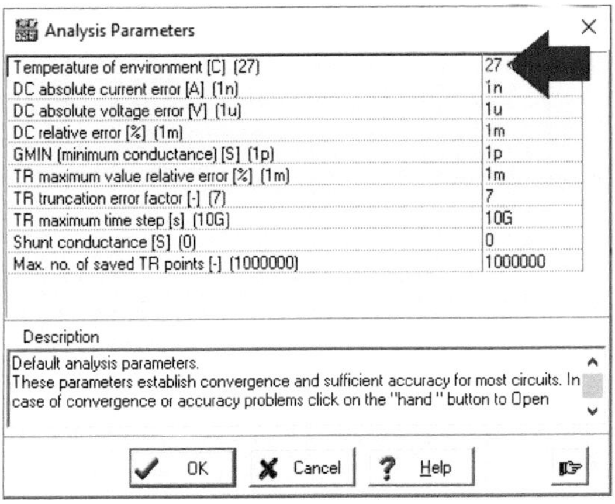

Figure 2.44

Temperature analysis permits you to study the effect of temperature on your circuit. Let's study an example. Consider the schematic shown in Figure 2.45. We want to sweep the temperature from 0 °C– 80 °C with 0.8 °C steps and see its effect on the voltage of the node 'out' (= voltage drop of the diode).

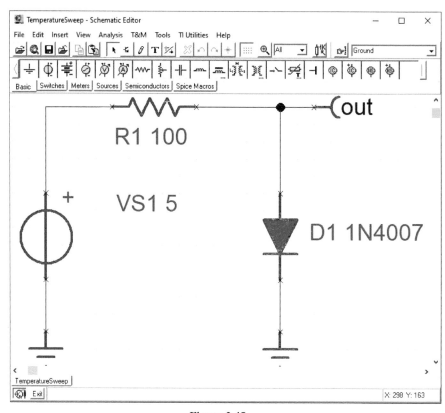

Figure 2.45

Click the Analysis > DC Analysis > Temperature Analysis (Figure 2.46). This opens the Temperature Analysis window Figure 2.47). Settings in Figure 2.47 sweep the temperature from 0 °C– 80 °C with 0.8 °C steps.

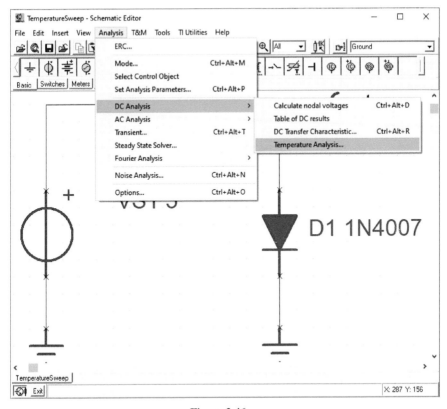

Figure 2.46

Figure 2.47

Click the OK button in Figure 2.47 to run the simulation. The simulation result is shown in Figure 2.48. According to the result shown in Figure 2.48, the voltage drops of the diode decreases as the temperature increases.

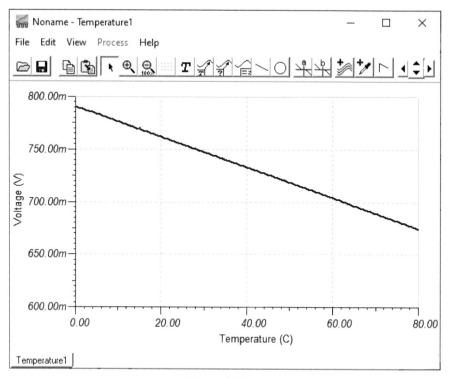

Figure 2.48

2.11 Example 10: Addition of SPICE Models to TINA-TI®

Sometimes the component that you want is not available in the TINA-TI. In this case, you can download the SPICE model of the component from the manufacturer website and import it to TINA-TI. The process of importing a SPICE model into TINA-TI is explained in this example.

Assume that you need the B340A diode. B340A is not available in TINA-TI. So, we need to search the internet in order to find its SPICE model (Figure 2.49).

Figure 2.49

The SPICE model of B340A is shown in Figure 2.50. This model must be imported into TINA-TI.

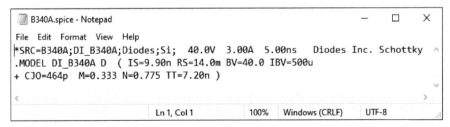

Figure 2.50

Click the Tools > New Macro Wizard (Figure 2.51). This opens the New Macro Wizard window (Figure 2.52).

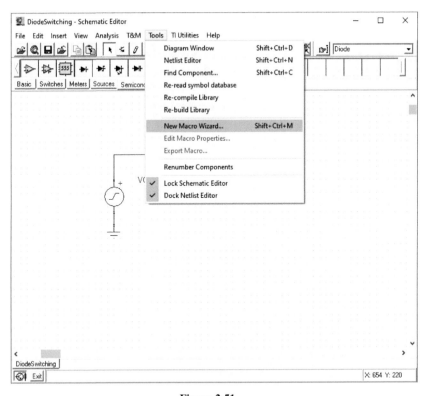

Figure 2.51

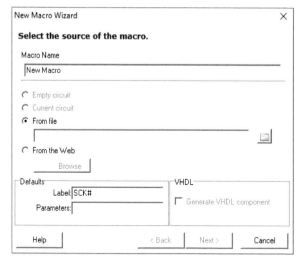

Figure 2.52

Enter the 'B340a' to the Macro Name box and use the open button to open the SPICE model of B340A (Figure 2.53). Then click the Next button. After clicking the Next button, the Save Macro window appears on the screen (Figure 2.54). Save the macro at the desired path. Importing the SPICE model into TINA-TI finishes at this point.

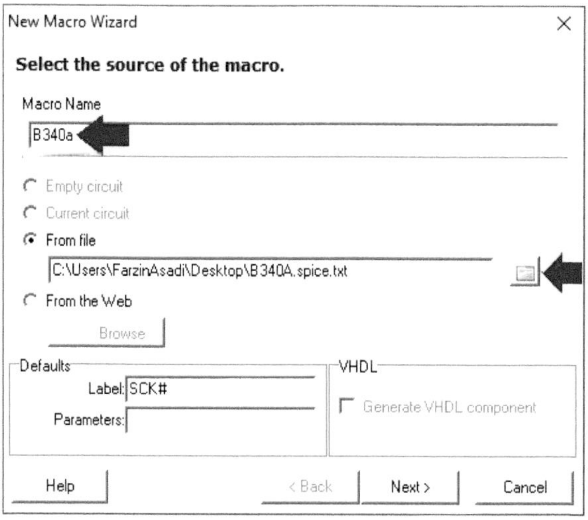

Figure 2.53

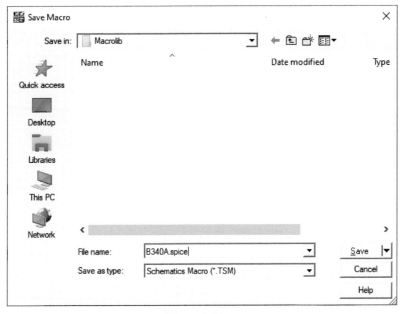

Figure 2.54

Let's add a B340A to the schematic. Click the Insert> Macro (Figure 2.55). This opens the Insert Macro window (Figure 2.56). Go to the path that you saved the macro (Figure 2.54) and open it. After opening the macro click on the schematic. This adds a B340A diode to your schematic (Figure 2.57).

Figure 2.55

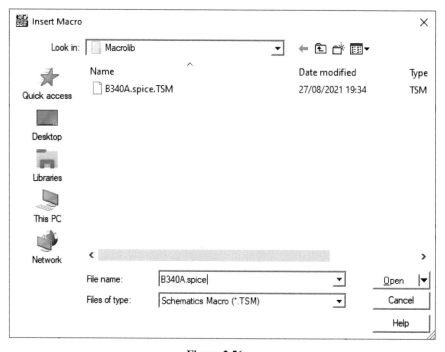

Figure 2.56

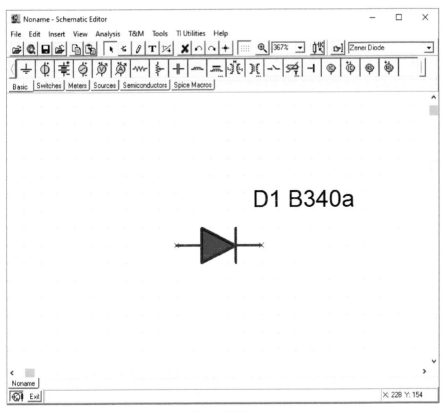

Figure 2.57

2.12 Example 11: Switching Behavior of Diodes

In this example, we will study the switching behavior of the B340A diode. Draw the schematic shown in Figure 2.58. Instead of using a B340A diode, you can use other Schottky diodes (Figure 2.59). Settings of Voltage Generator VG1 is shown in Figure 2.60.

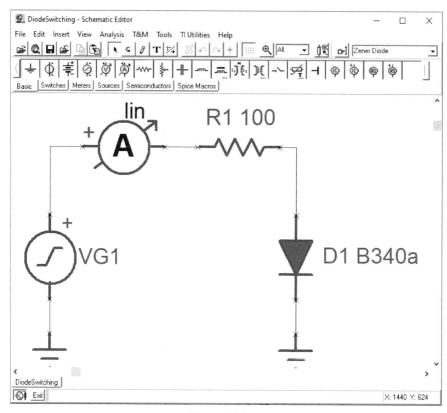

Figure 2.58

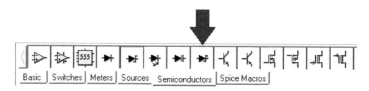

Figure 2.59

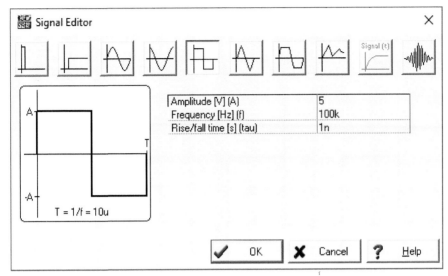

Figure 2.60

Run a transient analysis simulation with the parameters shown in Figure 2.61. The simulation results are shown in Figure 2.62.

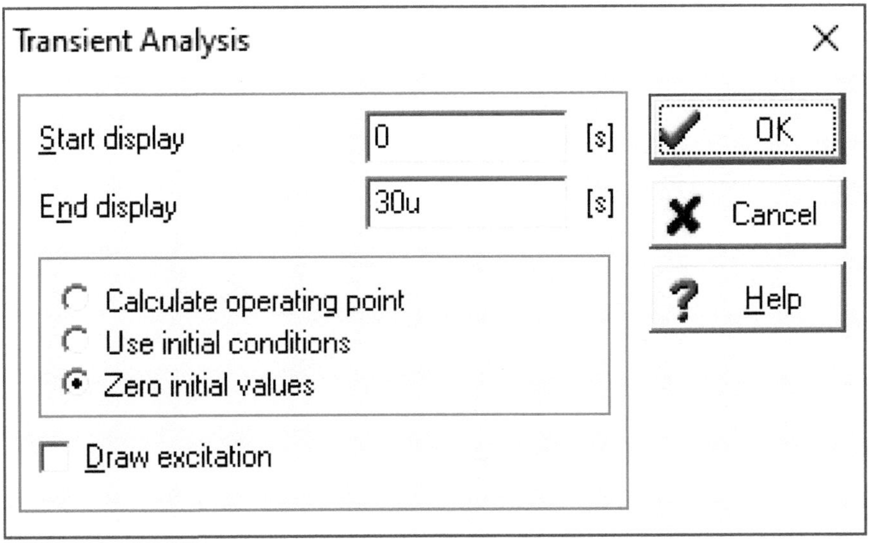

Figure 2.61

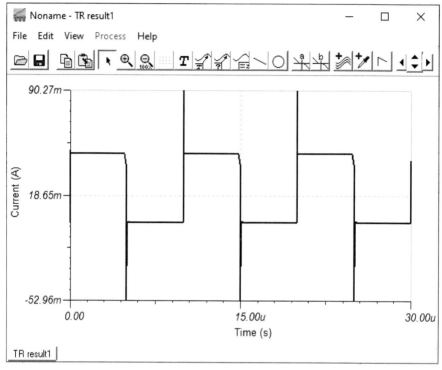

Figure 2.62

Use the magnifier icon to see the edges better. The rising edge (diode goes from off state to on state) is shown in Figure 2.63. According to Figure 2.64, the turn-on process took about 25.76 ns.

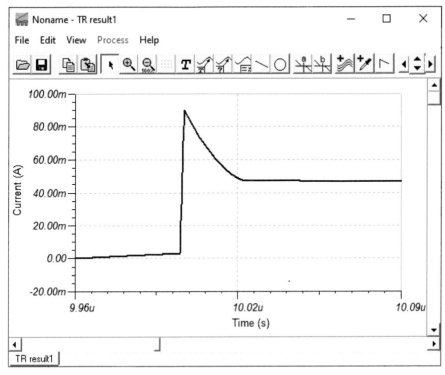

Figure 2.63

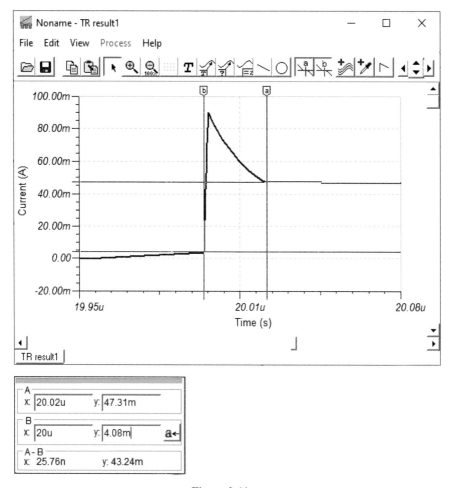

Figure 2.64

The falling edge (diode goes from one state to off state) is shown in Figure 2.65. According to Figure 2.66, the turn-off process took about 120.23 ns. Note that the turn-off process requires 4.67 times more time in comparison to the turn-on process.

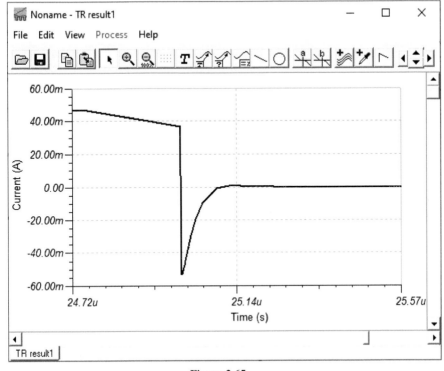

Figure 2.65

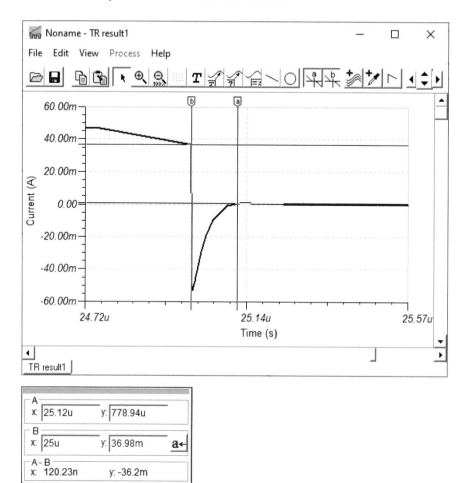

Figure 2.66

2.13 Example 12: Small Signal AC Resistance of Diodes

In this example, we want to measure the AC resistance of a diode. Draw the schematic shown in Figure 2.67.

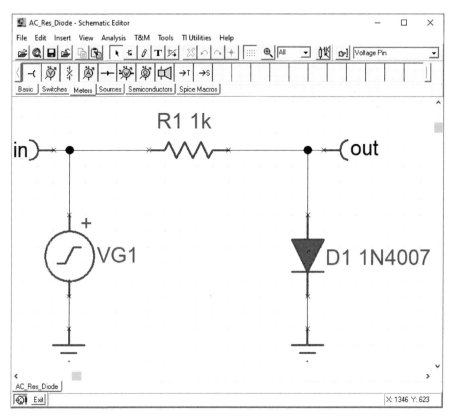

Figure 2.67

Settings of Voltage Generator VG1 are shown in Figure 2.68 and Figure 2.69. According to settings in Figure 2.68 and Figure 2.69 the input voltage is $1.65 + 0.1\sin(2\pi \times 50 \times t)\ V$.

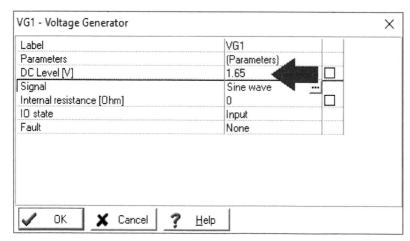

Figure 2.68

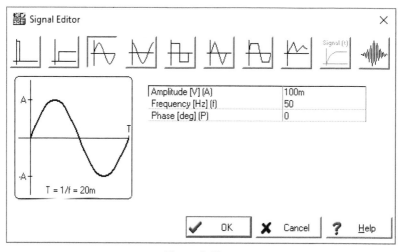

Figure 2.69

Run a transient analysis with the settings shown in Figure 2.70. The simulation result is shown in Figure 2.71.

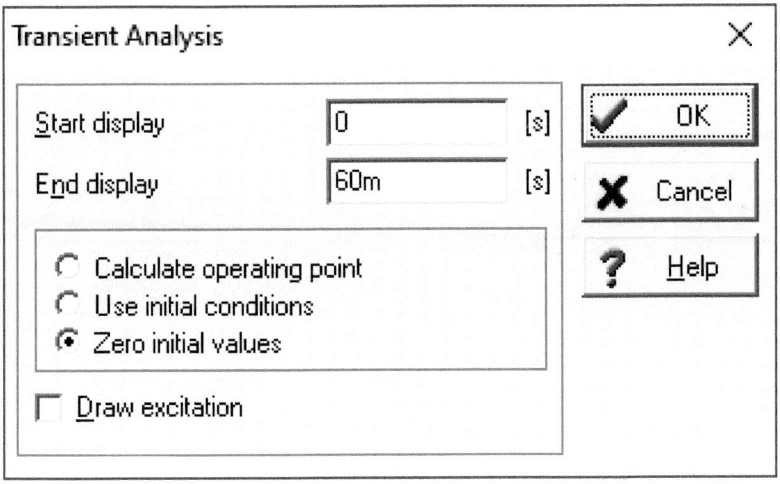

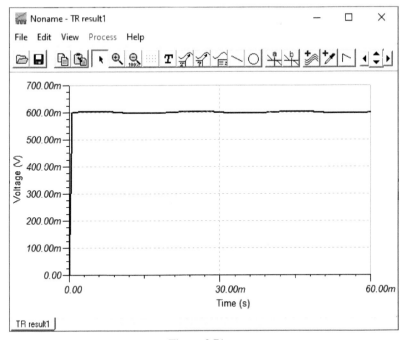

Figure 2.70

Figure 2.71

Use the magnifier icon to obtain a better view (Figure 2.72). According to Figure 2.73, peak of AC voltage at node 'out' is 7.44 mV.

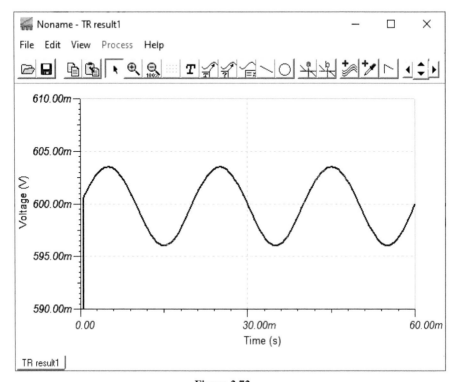

Figure 2.72

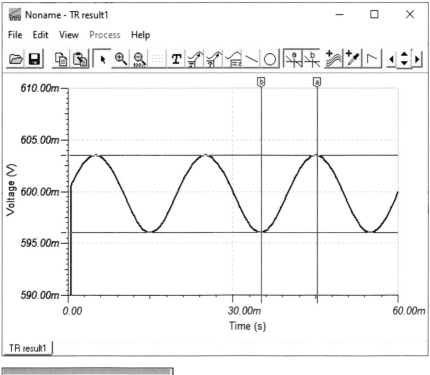

Figure 2.73

A small signal equivalent circuit is shown in Figure 2.74. Rac shows the small-signal AC resistance of the diode. Rac is unknown and it must be found. The MATLAB code shown in Figure 2.75 calculates the value of Rac. According to Figure 2.75, the value of Rac is 38.6373 Ω.

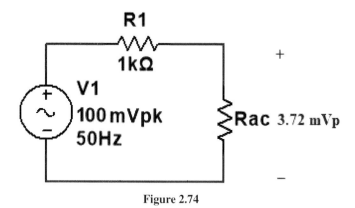

Figure 2.74

```
Command Window                                    ⊙
  >> R1=1e3;Vinp=100e-3;Vop=3.72e-3;
  >> syms Rd
  >> eval(solve(Rd/(Rd+R1)*Vinp==Vop))

  ans =

     38.6373

fx >> |
```

Figure 2.75

2.14 Example 13: Full Wave Rectifier (I)

In this example, we want to simulate a full-wave rectifier. Draw the schematic shown in Figure 2.76. Settings of Voltage Generator VG1 and transformer TR1 are shown in Figure 2.77 and Figure 2.78, respectively.

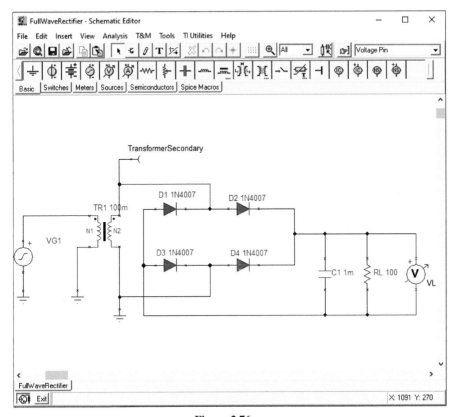

Figure 2.76

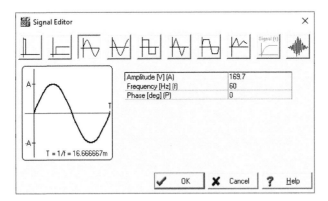

Figure 2.77

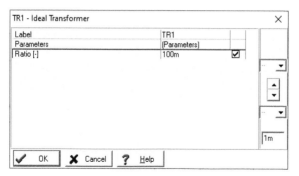

Figure 2.78

Click the Analysis > Transient (Figure 2.79). Run the transient analysis with the settings shown in Figure 2.80. The simulation result is shown in

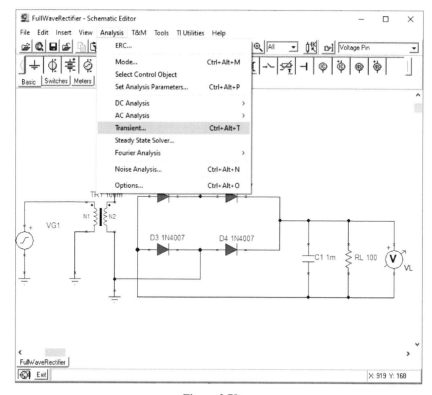

Figure 2.79

Figure 2.81. You can remove the sinusoidal signal (voltage of secondary of the transformer) by clicking on it and pressing the Delete key of the keyboard (Figure 2.82).

Figure 2.80

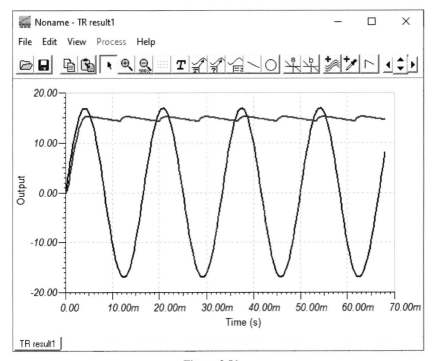

Figure 2.81

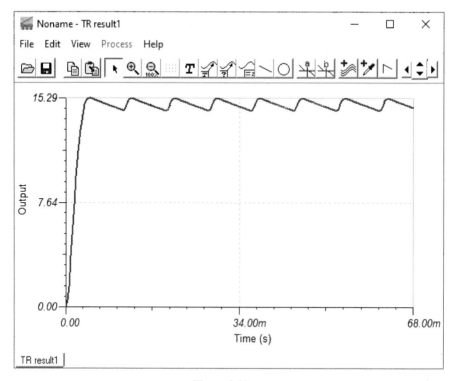

Figure 2.82

Let's do some measurement. According to Figure 2.83, the output voltage ripple is 920.13 mV.

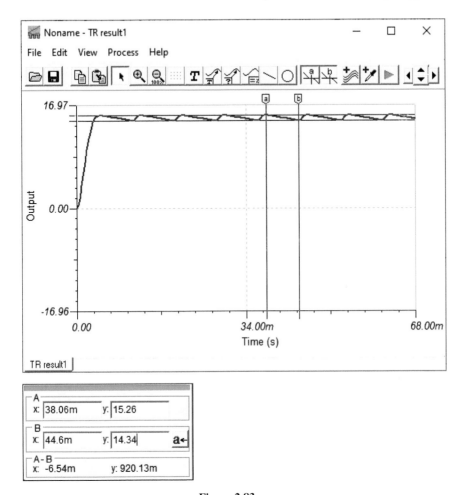

Figure 2.83

According to Figure 2.84, the frequency of the output voltage is $\frac{1}{|-8.36\ ms|} \approx 120$ Hz. Remember that in full-wave rectifiers, the frequency of output voltage is two times bigger than the frequency of the input AC source.

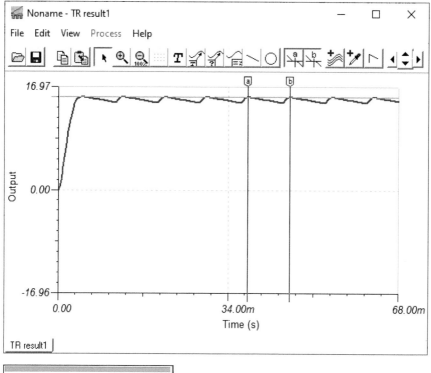

Figure 2.84

Let's measure the average and RMS values of the load voltage. Click on the load voltage waveform. Color of the waveform changes into red (Figure 2.85). Use the magnifier to select one cycle from the steady-state portion of the waveform (Figure 2.86). Then click the Process > Averages (Figure 2.87). This opens the Text window and shows the average and RMS values of the load voltage for you (Figure 2.88).

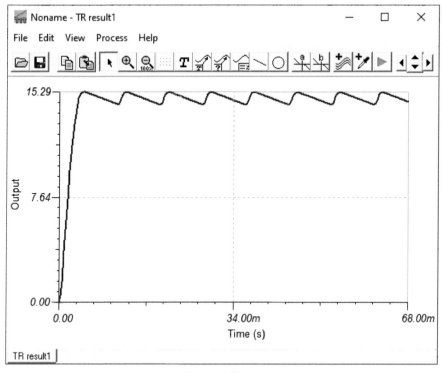

Figure 2.85

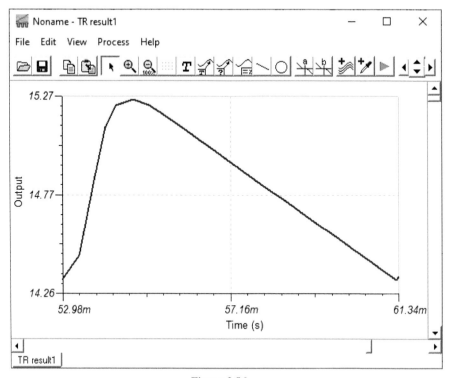

Figure 2.86

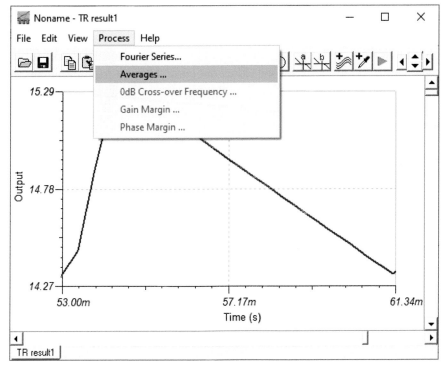

Figure 2.87

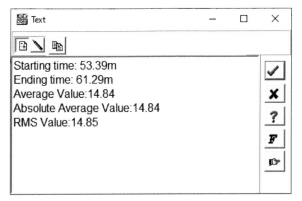

Figure 2.88

2.15 Example 14: Full Wave Rectifier (II)

In this example, we want to observe the current passed from diodes and the current drawn from the input AC source. Add two Ampere meter blocks to the schematic (Figure 2.89). Iin measures the current drawn from the input AC source and ID2 measures the current passed from the diode D2.

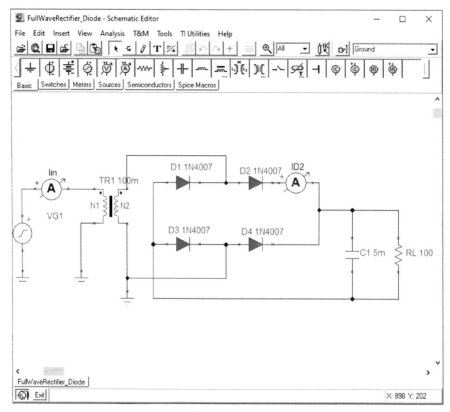

Figure 2.89

Run a transient analysis with the parameters shown in Figure 2.90. The simulation result is shown in Figure 2.91. Both currents (the current drawn from the input source and diode D2's current) are shown on the screen.

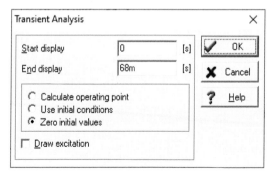

Figure 2.90

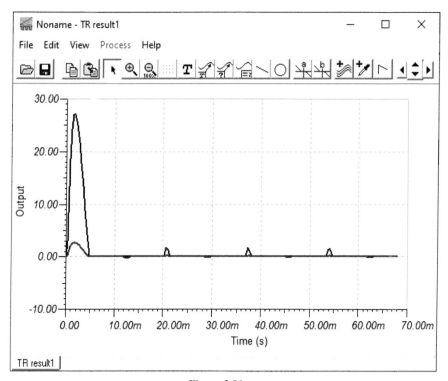

Figure 2.91

Click the View > Show/Hide curves (Figure 2.92). This opens the Show/Hide curves window (Figure 2.93).

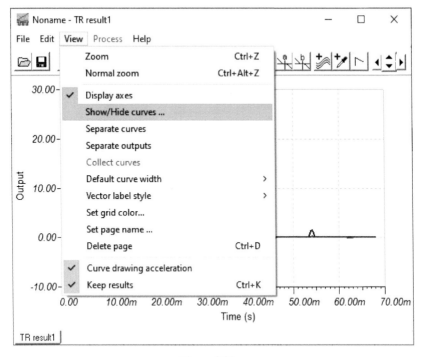

Figure 2.92

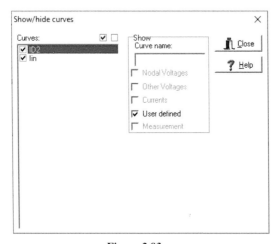

Figure 2.93

Uncheck the ID2 (Figure 2.94) and click the Close button. This hides the current passed from the diode D2 (Figure 2.95).

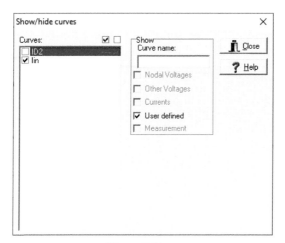

Figure 2.94

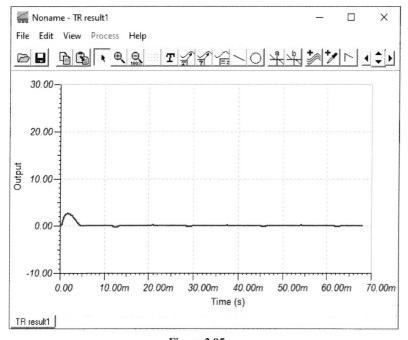

Figure 2.95

Right-click on the vertical axis and click the Properties (Figure 2.96). This opens the Set Axis window. Do the settings similar to Figure 2.97 and click the OK button. After clicking the OK button, the graph changes to what is shown in Figure 2.98. Now you can see the details better. The maximum of this waveform is 2.71 A.

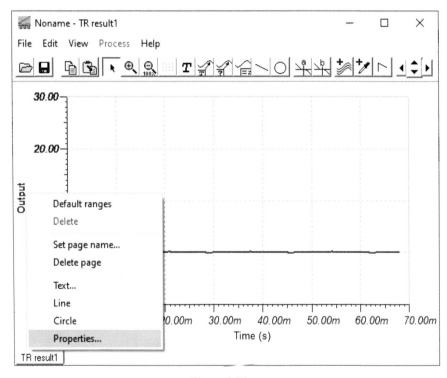

Figure 2.96

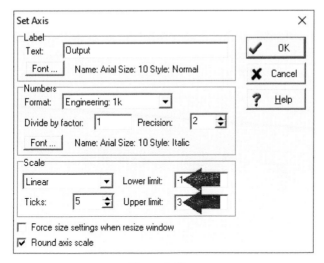

Figure 2.97

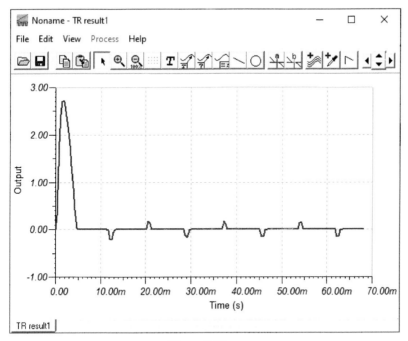

Figure 2.98

Use the magnifier to see the steady-state portion of the waveform. As shown in Figure 2.99, the current drawn from the input AC source is a train of narrow pulses. Such narrow pulses contain many harmonics. Therefore, we expect a low power factor for such a rectifier.

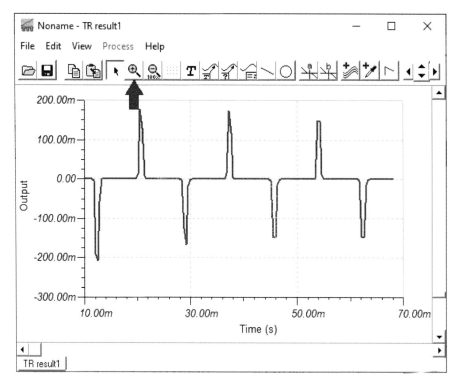

Figure 2.99

Click the View> Show/Hide curves (Figure 2.92). This opens the Show/Hide curves window. Uncheck the Iin box and check the ID2 box (Figure 2.100). Now, the current passed from the diode appears on the screen (Figure 2.101). The maximum current passed from the diode D2 is 27.06 A. This maximum value must be less than the non-repetitive peak forward surge current (I_{FSM}) value in the data sheet.

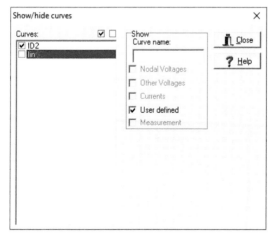

Figure 2.100

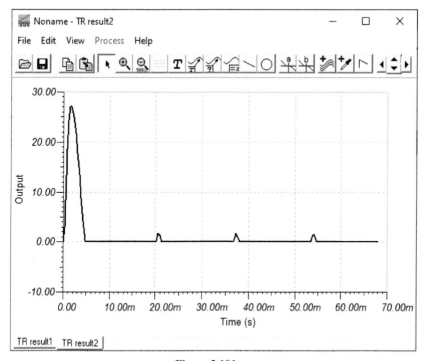

Figure 2.101

Use the magnifier icon to see the steady-state portion of the graph (Figure 2.102). The current passed through diodes are short pulses with a peak value of about 1.8 A.

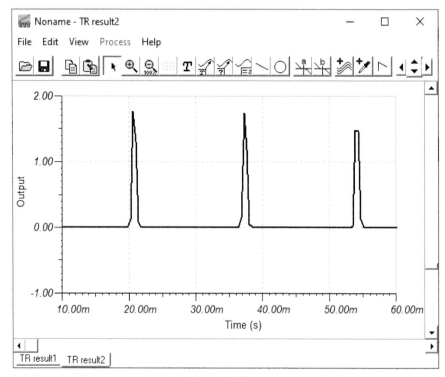

Figure 2.102

Let's measure the average value of the waveform shown in Figure 2.102. Use magnifier icon to select one cycle from steady-state portion of the waveform (Figure 2.103).

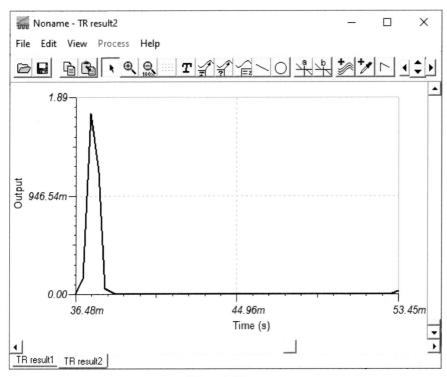

Figure 2.103

Click on the waveform to select it. After clicking on the waveform, the color of graph changes into red (Figure 2.104).

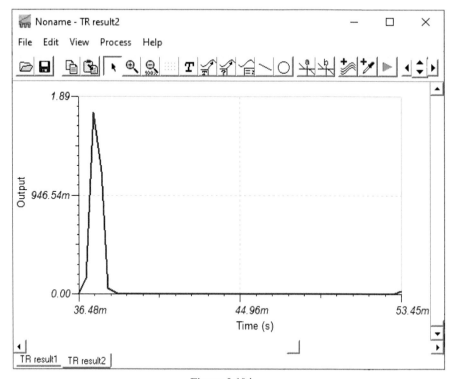

Figure 2.104

Click the Process > Averages (Figure 2.105). This opens the Text window. According to Figure 2.106, an average value of current is 74.09 mA. This value must be less than the maximum average forward rectified current value in the data sheet.

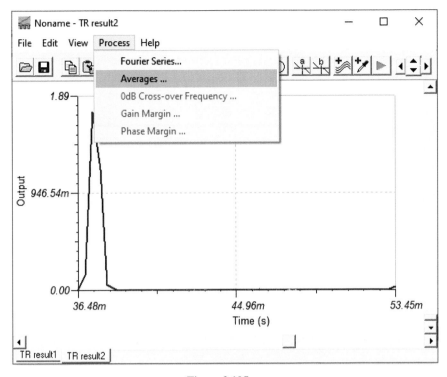

Figure 2.105

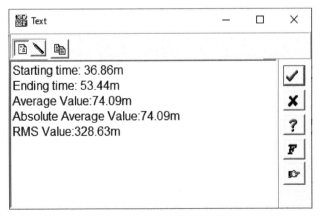

Figure 2.106

2.16 Example 15: Controlled Rectifier

The output voltage of diode rectifiers is constant. If you need to control the output voltage, you need to use controlled rectifiers. Controlled rectifiers use one or more thyristors (SCR). TINA-TI doesn't have any ready to use SCR. Let's download the SCR models from the internet and add it to TINA-TI. Go to this page: https://bit.ly/3DBIkmo (Figure 2.107)

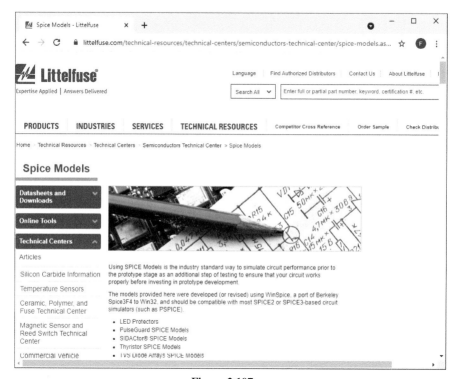

Figure 2.107

Click the Thyristor SPICE Models (Figure 2.108) and download the SPICE_Littelfuse_SCR-EC103xx-SxSx_A.lib (Figure 2.109).

Using SPICE Models is the industry standard way to simulate circuit performance prior to the prototype stage as an additional step of testing to ensure that your circuit works properly before investing in prototype development.

The models provided here were developed (or revised) using WinSpice, a port of Berkeley Spice3F4 to Win32, and should be compatible with most SPICE2 or SPICE3-based circuit simulators (such as PSPICE).

- LED Protectors
- PulseGuard SPICE Models
- SIDACtor® SPICE Models
- Thyristor SPICE Models
- TVS Diode Arrays SPICE Models
- TVS Diode SPICE Models
- Varistor SPICE Models

Figure 2.108

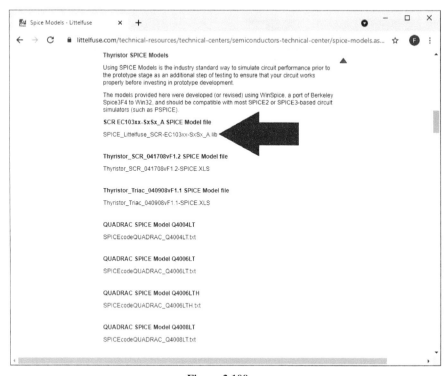

Figure 2.109

You can open the downloaded library file with Notepad (Figure 2.110). Part number of modeled thyristors are written in front of the SUBCKT lines. 16 thyristors are modelled in this file (Figure 2.111).

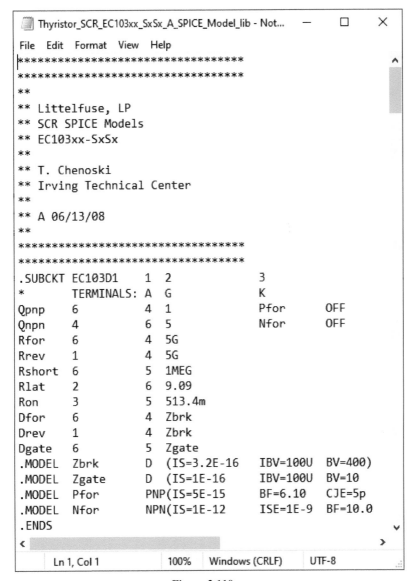

Figure 2.110

.SUBCKT EC103D1	.SUBCKT S4S1
.SUBCKT EC103M1	.SUBCKT S6S1
.SUBCKT EC103D2	.SUBCKT S4S2
.SUBCKT EC103M2	.SUBCKT S6S2
.SUBCKT EC103D	.SUBCKT S4S
.SUBCKT EC103M	.SUBCKT S6S
.SUBCKT EC103D3	.SUBCKT S4S3
.SUBCKT EC103M3	.SUBCKT S6S3

Figure 2.111

Let's add the EC103D1 to TINA-TI. Open the TINA-TI and click the Tools > New Macro Wizard or press the Ctrl+Shift+M. This opens the New Macro Wizard window. Enter EC103D1 to the Macro Name box and open the downloaded library file (Figure 2.112).

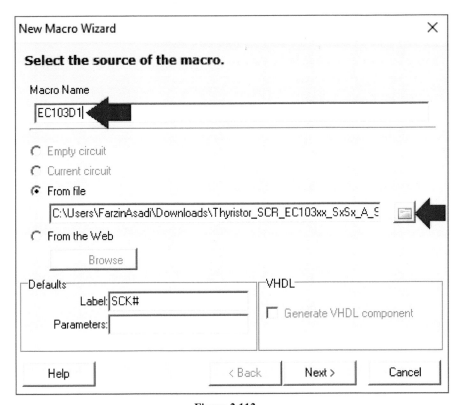

Figure 2.112

Click the Next button in Figure 2.112. This opens the New Macro Wizard window (Figure 2.113). Select the EC103D1

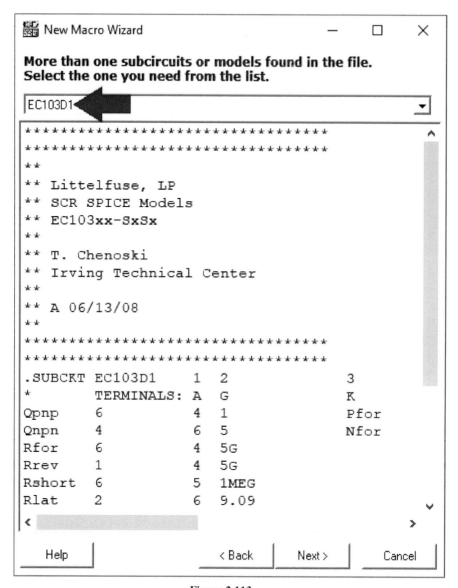

Figure 2.113

After clicking the Next button in Figure 2.113, the window shown in figure Figure 2.114 appears on the screen. TINA-TI suggests the symbol shown in Figure 2.114 to the new part. Let's select a better symbol for it.

Change the settings to what is shown in Figure 2.115 and click the Next button.

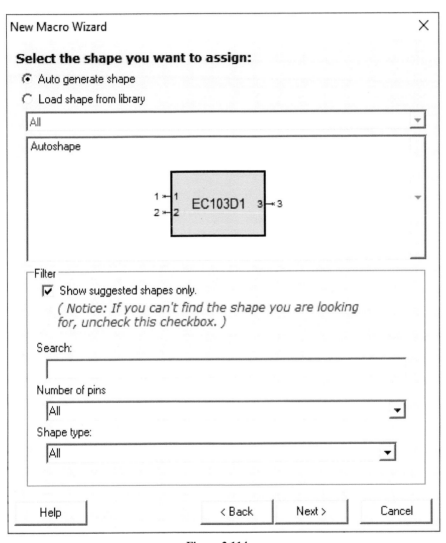

Figure 2.114

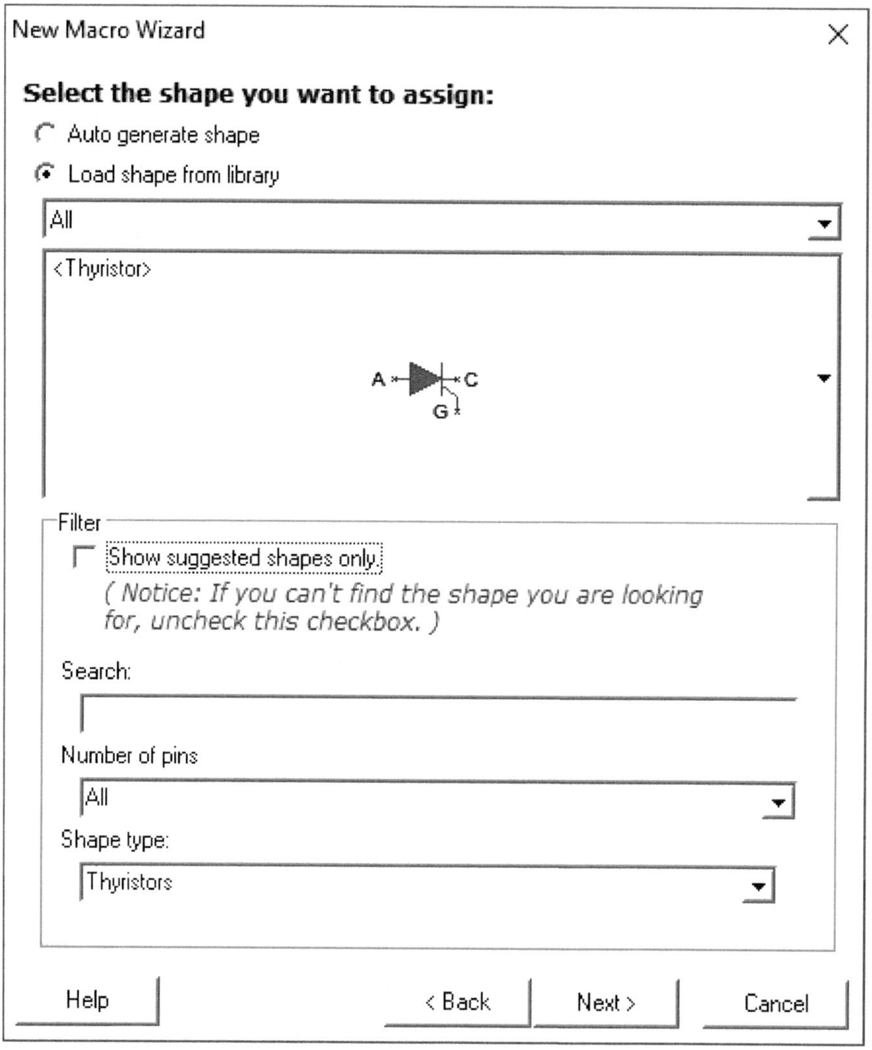

New Macro Wizard   ✕

Select the shape you want to assign:

○ Auto generate shape

◉ Load shape from library

All   ▼

<Thyristor>

A ▸▶—⊣⊦◂ C
     G

▼

Filter

☐ Show suggested shapes only.

(*Notice: If you can't find the shape you are looking for, uncheck this checkbox.*)

Search:

Number of pins

All   ▼

Shape type:

Thyristors   ▼

Help   < Back   Next >   Cancel

Figure 2.115

After clicking the Next button in Figure 2.115, the window is shown in the Figure 2.116 appears on the screen. According to the SUBCKT line shown in Figure 2.116, Anode is 1, Gate is 2 and Cathode is 3. Therefore, you need to drag and drop 1 onto the Anode terminal, drag and drop 2 onto the Gate terminal and drag and drop 3 onto the Cathode terminal (Figure 2.117).

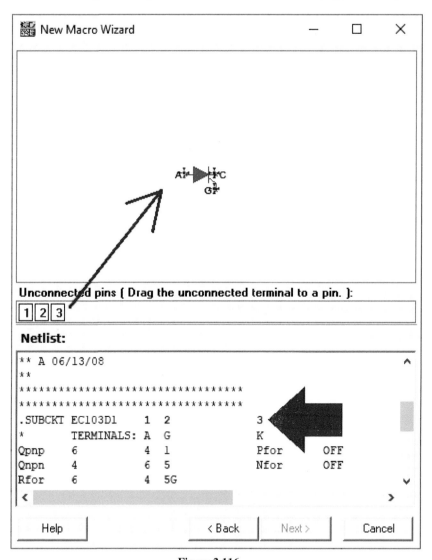

Figure 2.116

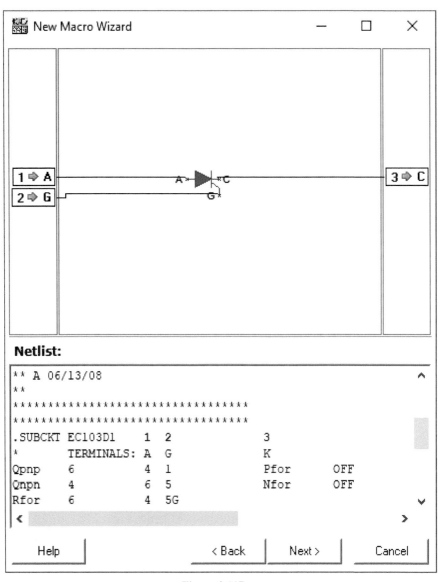

Figure 2.117

After clicking the Next button in Figure 2.117, the window shown in Figure 2.118 appears on the screen. Save the macro at the desired path. After saving the macro, the message shown in Figure 2.119 appears on the screen. Click the Finish button.

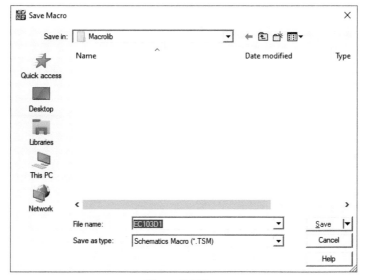

Figure 2.118

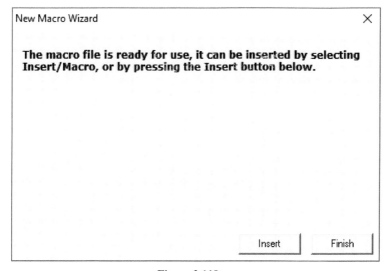

Figure 2.119

Let's simulate a simply controlled rectifier with the imported thyristor. Draw the schematic shown in Figure 2.120. Settings of input Voltage Generator VG1 are shown in Figure 2.121.

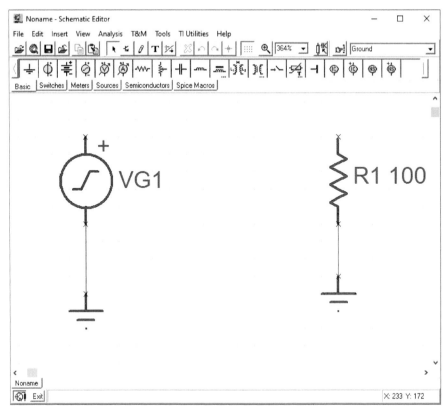

Figure 2.120

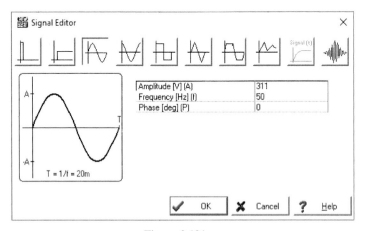

Figure 2.121

It's time to add the thyristor. Click the Insert > Macro (Figure 2.122).

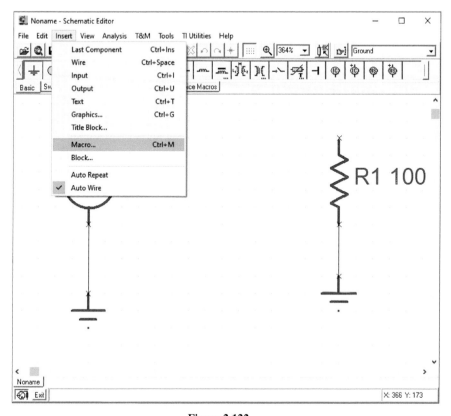

Figure 2.122

Go to the path that you saved the macro (Figure 2.118) and open it (Figure 2.123). After opening the macro, click on the schematic and this adds an EC103D1 thyristor to the schematic (Figure 2.124).

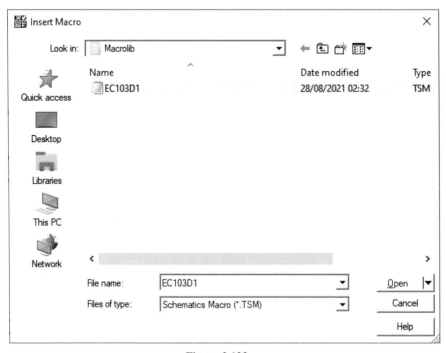

Figure 2.123

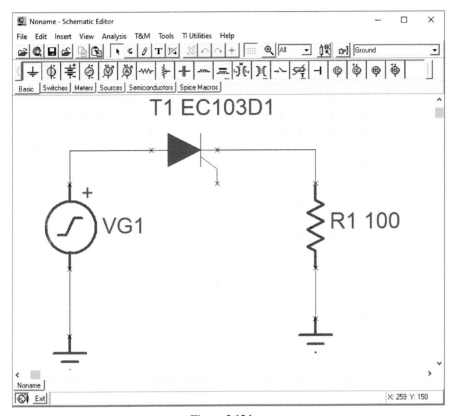

Figure 2.124

Add a Voltage Generator VG2 and resistor R2 to the schematic (Figure 2.125). Settings of Voltage Generator VG2 is shown in Figure 2.126 and Figure 2.127. The output of the Voltage Generator VG2 is shown in the Figure 2.128 (A=4.99 ms, B=5 ms, C=8 ms, D=8.01 ms and E=20 ms). So, the thyristor is triggered at t=5 ms and the firing angle is $\frac{5 \text{ ms}}{20 \text{ ms}} \times 360° = 90°$.

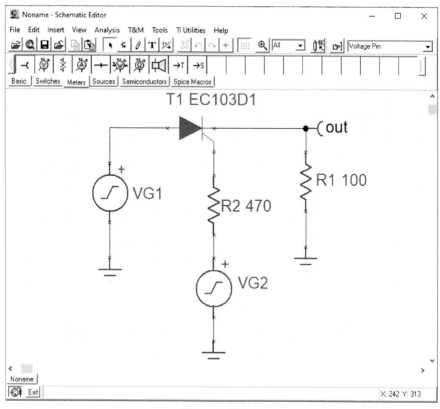

Figure 2.125

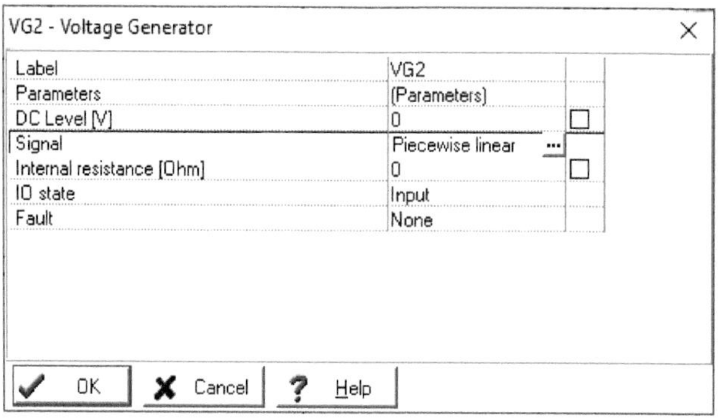

Figure 2.126

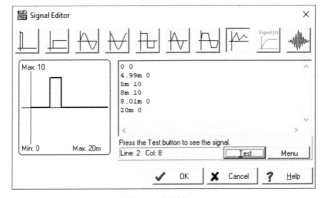

Figure 2.127

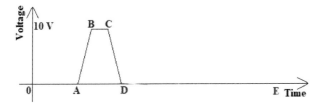

Figure 2.128

Run a transient analysis with the parameters shown in Figure 2.129. The simulation result is shown in Figure 2.130.

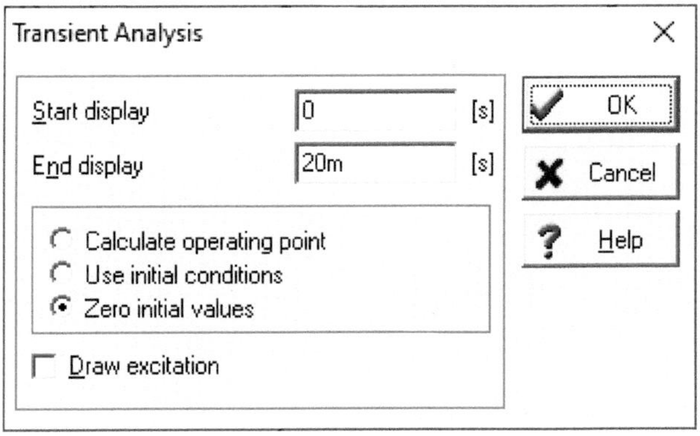

Figure 2.129

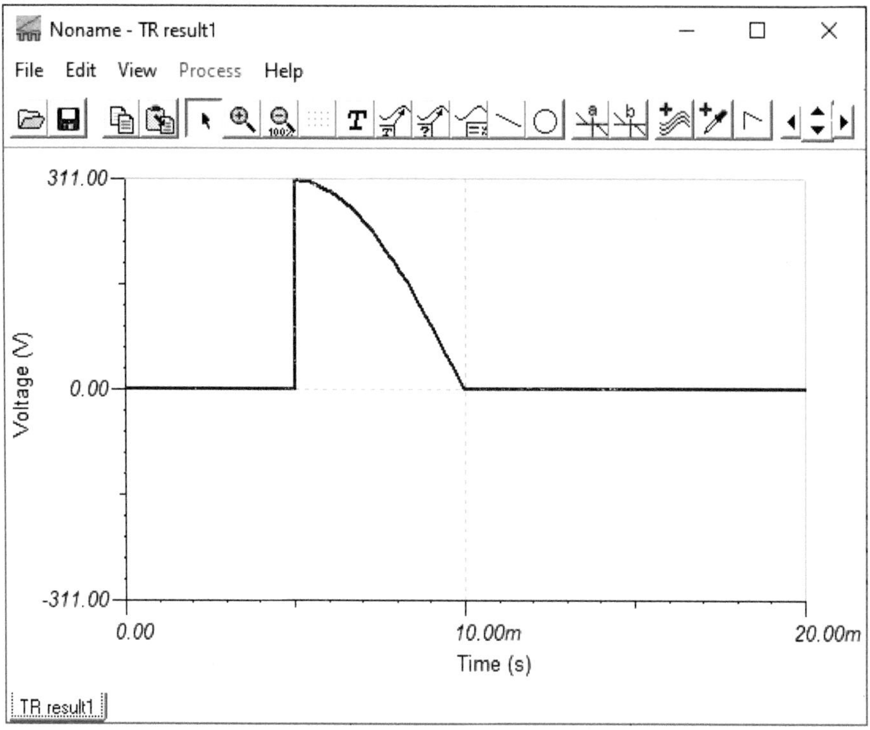

Figure 2.130

Use a cursor to measure the maximum of an obtained waveform. According to Figure 2.131, the maximum of this waveform is 308.21 V. So, the voltage drop of the thyristor is 311–308.21=2.79 V.

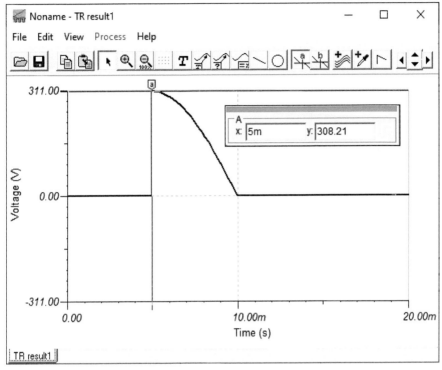

Figure 2.131

Let's measure the average and RMS values of the output waveform. Click on the waveform to select it (Figure 2.132) and click the Process > Averages. According to Figure 2.133, the average and RMS value of this waveform is 49.05 V and 109.01 V, respectively.

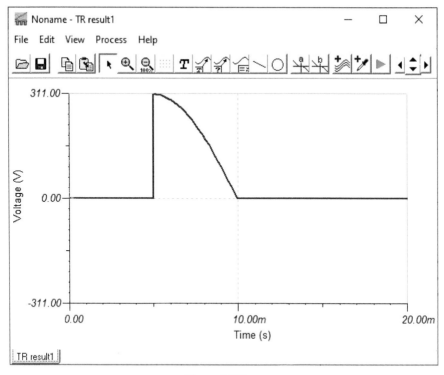

Figure 2.132

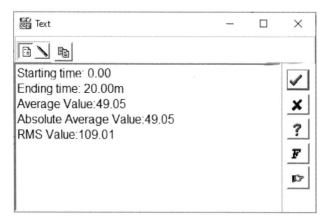

Figure 2.133

Let's check the obtained results. The MATLAB code shown in Figure 2.134 calculates the average and RMS values of the waveform shown in Figure 2.132. Obtained results are quite close to TINA-TI results shown in Figure 2.133.

```
Command Window

    >> syms t
    >> V=308.21*sin(2*pi*50*t);
    >> T=1/50;
    >> Ave=eval(1/T*int(V,5e-3,10e-3))

  Ave =

      49.0531

    >> RMS=eval(sqrt(1/T*int(V^2,5e-3,10e-3)))

  RMS =

      108.9687

fx >> |
```

Figure 2.134

2.17 Example 16: Measurement of Operating Point of Common Emitter Amplifier

In this example, we want to simulate the common emitter amplifier shown in Figure 2.135. The transistor block can be found in the Semiconductors tab (Figure 2.136).

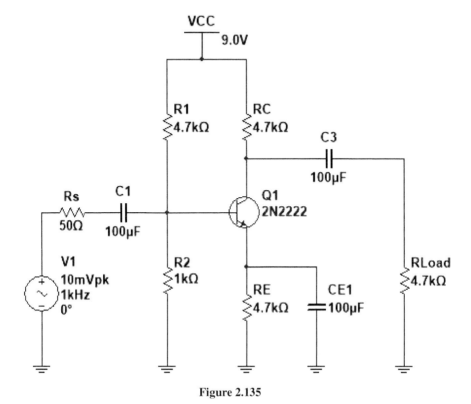

Figure 2.135

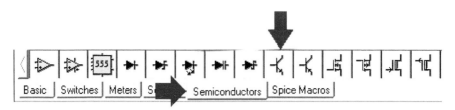

Figure 2.136

Draw the schematic shown in Figure 2.137. Settings of Voltage Generator VG1 is shown in Figure 2.138. VCC are a Jumper block (Figure 2.139).

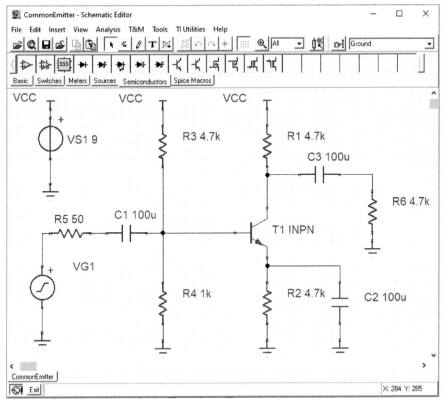

Figure 2.137

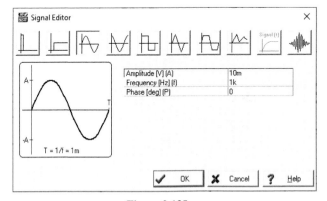

Figure 2.138

Figure 2.139

Double click the transistor block. This opens the NPN Bipolar Transistor window (Figure 2.140). Click the three dots behind the Type box (Figure 2.140). This opens the Catalog Editor window (Figure 2.141). Now you can select the part that you want.

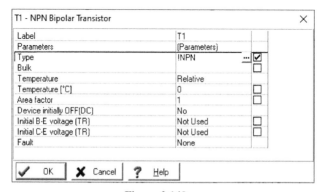

Figure 2.140

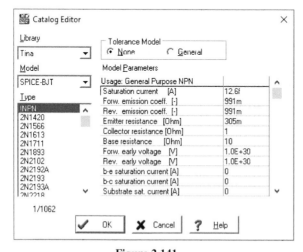

Figure 2.141

Select the 2N2222 and click the OK button (Figure 2.142). After clicking the OK button in Figure 2.142, the schematic changes to what is shown in Figure 2.143.

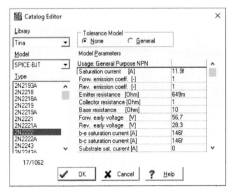

Figure 2.142

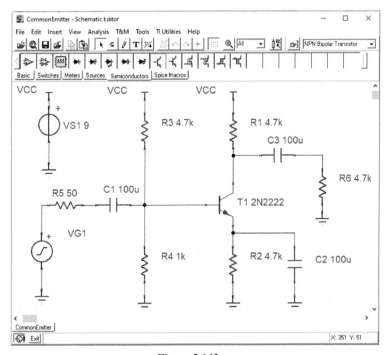

Figure 2.143

Let's do a DC analysis and measure the operating point of the transistor. Click the Analysis > DC Analysis> Calculate nodal voltages (Figure 2.144).

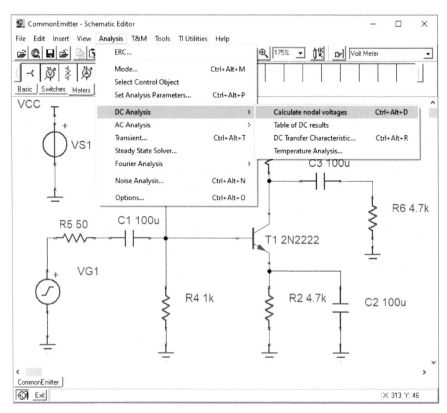

Figure 2.144

After clicking the Calculate nodal voltages, the mouse pointer changes into a probe. Click on the base, collector and emitter nodes to read their voltages. Values of the base, collect and emitter voltages are shown in Figure 2.145, Figure 2.146 and Figure 2.147, respectively. So, collector-emitter voltage (VCE) is 8.04–0.96994= 7.07 V.

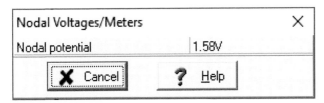

Figure 2.145

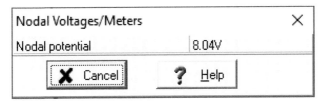

Figure 2.146

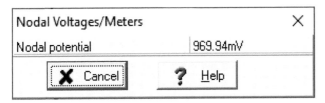

Figure 2.147

Let's measure the base and collector currents. Add two Ampere meter blocks to the schematic (Figure 2.148) and click the Analysis > DC Analysis> Calculate nodal voltages (Figure 2.144). The simulation result is shown in the Figure 2.149. The collector current is 204.48μA. So, the operating point of the transistor is (7.07 V, 0.20448 mA). The current gain of a transistor is $\beta = \frac{I_C}{I_B} = \frac{204.48 \ \mu A}{1.89 \ \mu A} = 108.19$ at this operating point.

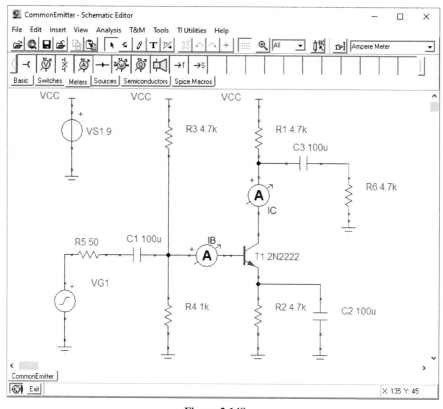

Figure 2.148

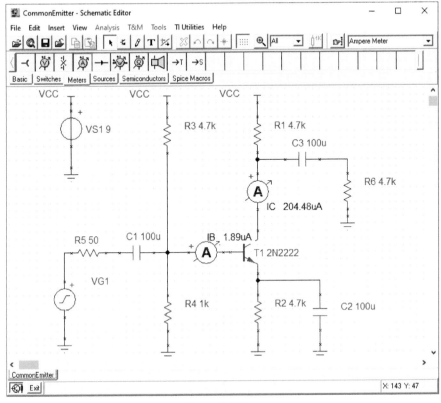

Figure 2.149

2.18 Example 17: Measurement of Voltage Gain for Common Emitter Amplifier

In this example, we want to measure the voltage gain of the common emitter amplifier of Example 16. Let's start. Change the schematic of Example 16 to what is shown in the Figure 2.150.

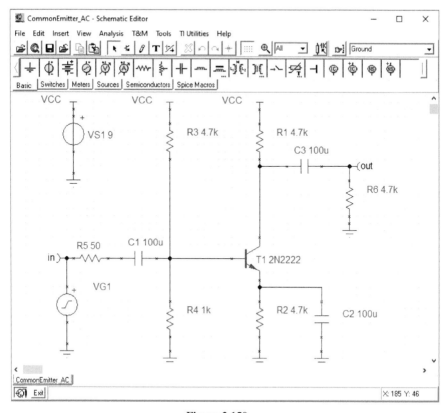

Figure 2.150

Click the Analysis > AC Analysis > Calculate nodal voltages (Figure 2.151). The result is shown in Figure 2.152. According to Figure 2.152, the RMS of the input voltage and output voltage is 7.07 mV and 121.54 mV, respectively. Therefore, the voltage gain of an amplifier is $\frac{121.54\ mV}{7.07\ mV} = 17.19$.

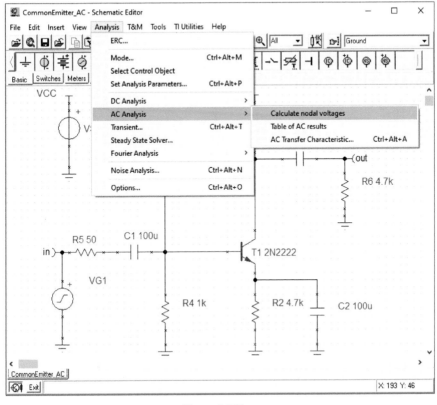

Figure 2.151

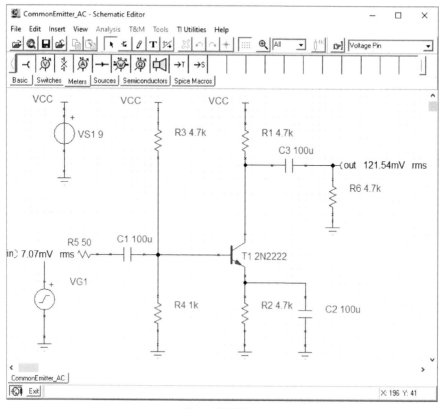

Figure 2.152

2.19 Example 18: Total Harmonic Distortion (THD) of Common Emitter Amplifier (I)

In this example we want to measure the THD of common emitter amplifier of Example 16. Open the schematic of Example 16 (Figure 2.153).

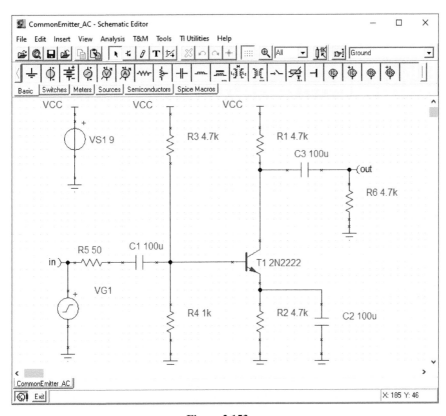

Figure 2.153

Click the Analysis > Transient (Figure 2.154).

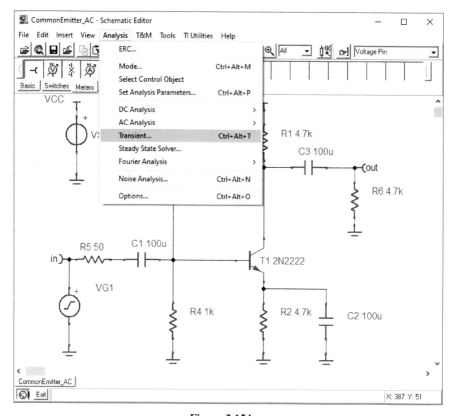

Figure 2.154

Run a transient analysis with the parameters shown in Figure 2.155. The simulation result is shown in Figure 2.156.

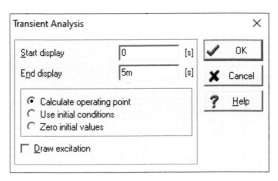

Figure 2.155

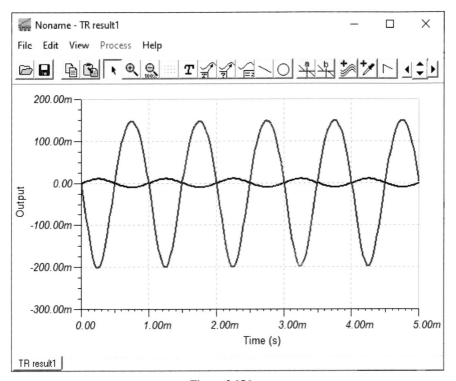

Figure 2.156

Note that the output waveform is in the Figure 2.156 is distorted (note that the output waveform is not symmetric). According to figure 2.157, the positive and negative peaks are at ±148.88 mV and −196.83 mV, respectively.

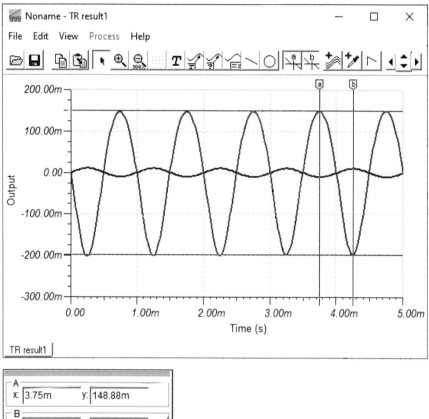

Figure 2.157

Let's measure the THD of the output voltage waveform. Right-click on the output voltage waveform and click the Fourier Series (Figure 2.158). This opens the Fourier Series window (Figure 2.159). Click the Calculate button to start the calculations. The result is shown in Figure 2.160. According to figure 2.160, THD is 8.6995%.

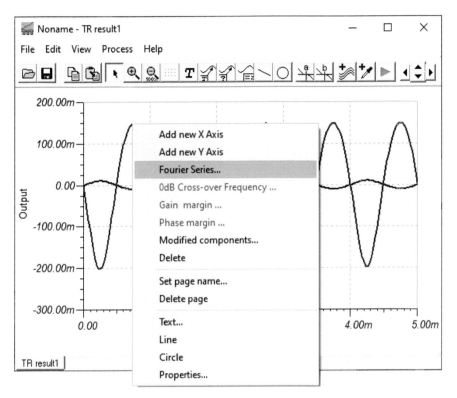

Figure 2.158

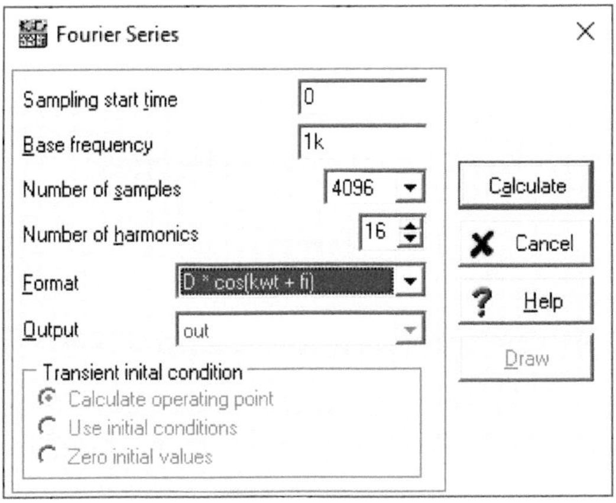

Figure 2.159

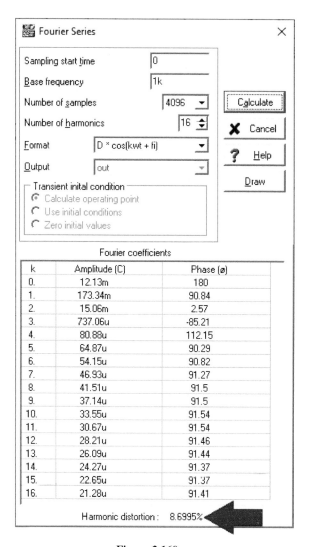

Figure 2.160

2.20 Example 19: THD of Common Emitter Amplifier (II)

The THD of an amplifier can be calculated with the aid of Fourier Analysis as well. In this example, we will use this method to measure the output THD of the amplifier of Example 16. Let's start. Open the schematic of Example 16 (Figure 2.161).

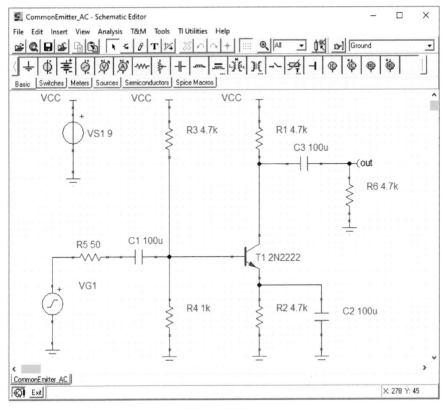

Figure 2.161

Click the Analysis> Fourier Analysis > Fourier Series (Figure 2.162). This opens the Fourier Series window. Click the Calculate button. According to Figure 2.163, the THD of the amplifier is 8.7487%. The obtained result is quite close to the result of the previous example.

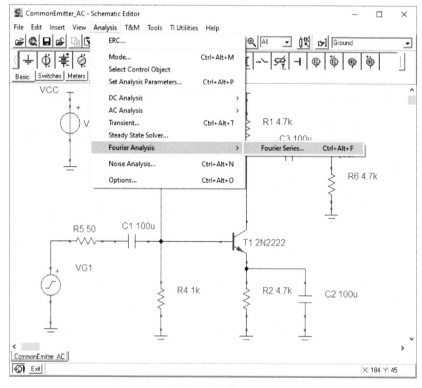

Figure 2.162

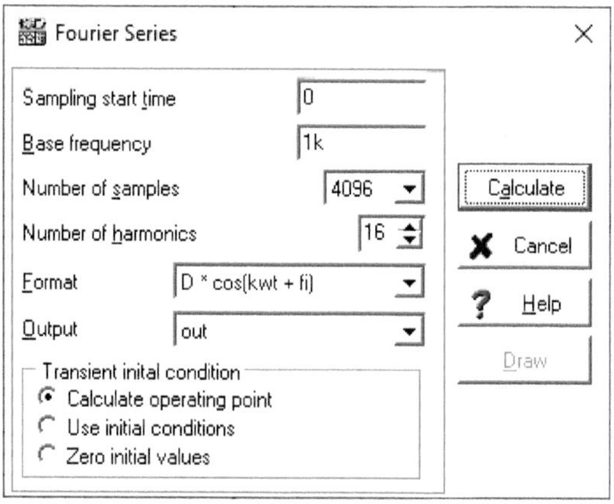

Figure 2.163

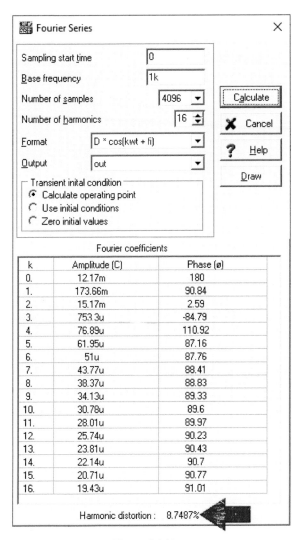

Figure 2.164

2.21 Example 20: Frequency Response of Common Emitter Amplifier (I)

In this example, we want to find the frequency response of the common emitter amplifier of Example 16. Let's start. Open the schematic of Example 16 and add a Voltage Pin block to it (Figure 2.165).

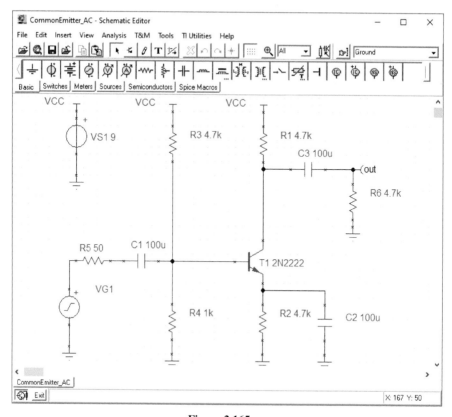

Figure 2.165

Click the Analysis > AC Analysis > AC Transfer Characteristic (Figure 2.166) and run the simulation with the settings shown in figure Figure 2.167. These settings calculate the frequency response ($\frac{V_{out}(j\omega)}{VG1(j\omega)}$) on the [10 Hz, 100 MHz] range.

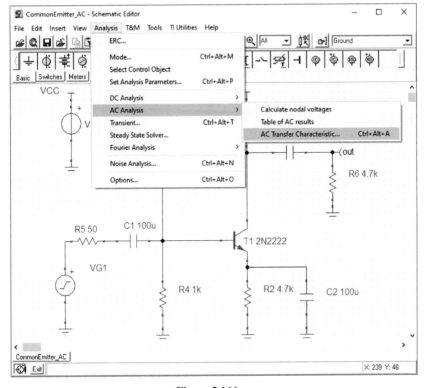

Figure 2.166

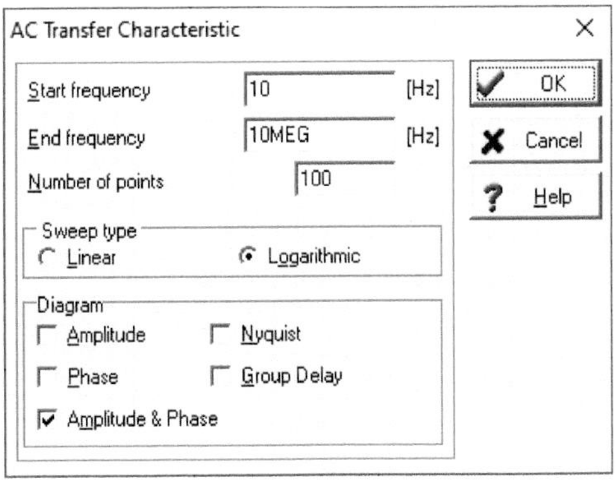

Figure 2.167

After clicking the OK button in Figure 2.167, the result shown in Figure 2.168 appears on the screen.

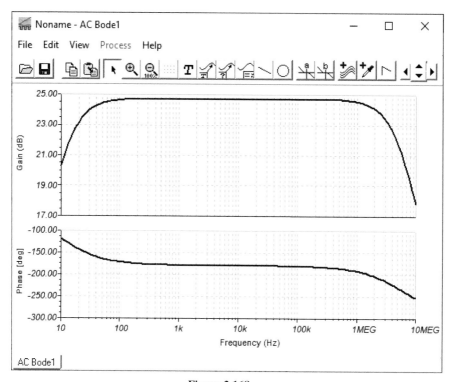

Figure 2.168

Let's measure the mid-band gain of the system. According to Figure 2.169, the mid-band gain is 24.71 dB. 24.71 dB equals to the normal gain of $10^{\frac{24.71}{20}} = 17.1989$.

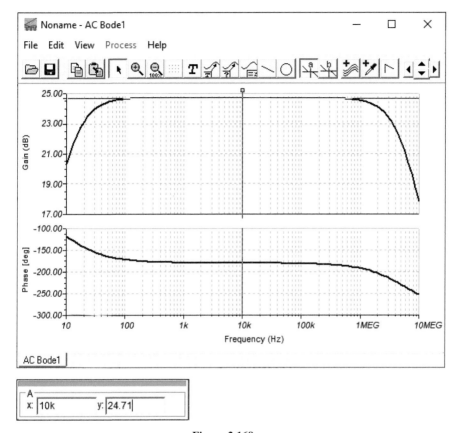

Figure 2.169

Let's measure the bandwidth of the system. We need to find the −3dB points. Mid band gain is 24.71 dB. So, we need to find the points with the gain of 24.71−3=21.71 dB. According to Figure 2.170 and Figure 2.171, these frequencies are 13.17 Hz and 5.1 MHz. Therefore, the bandwidth is 5.1 MHz − 13.17 Hz≈5.1 MHz. It is quite difficult to obtain such a high bandwidth with a single-stage amplifier. One important reason is the presence of stray capacitors. This subject is studied in the next example.

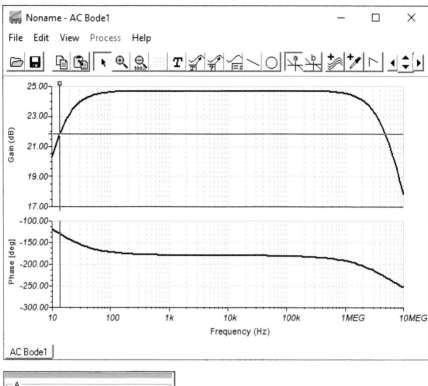

Figure 2.170

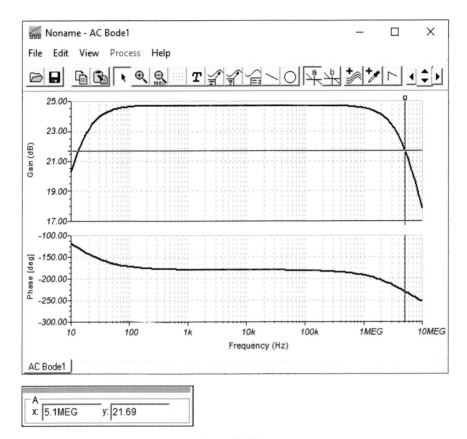

Figure 2.171

2.22 Example 21: Frequency Response of Common Emitter Amplifier (II)

In this example, we want to study the effect of stray capacitors on the frequency response of the amplifier. Open the schematic of Example 20 and add a 100 pF capacitor between the base and collector terminals (Figure 2.172).

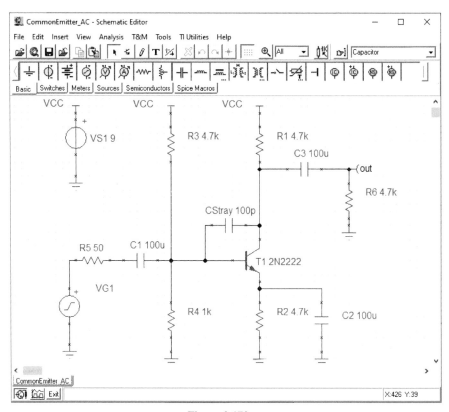

Figure 2.172

Click the Analysis> AC Analysis> AC Transfer Characteristic (Figure 2.166) and run the simulation with the settings shown in Figure 2.167. The simulation result is shown in Figure 2.173.

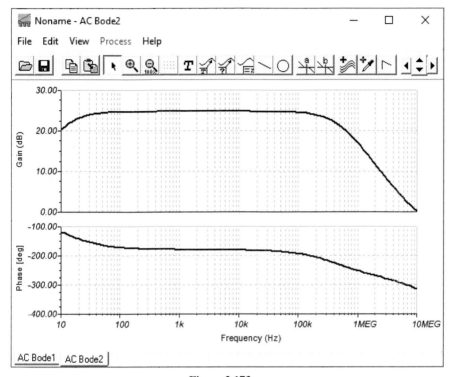

Figure 2.173

Let's measure the mid-band gain. According to Figure 2.174, the mid-band gain is 24.7 dB. The mid-band gain of the previous example was 24.71 dB. So, the 'CStray' didn't affect the mid band gain of an amplifier.

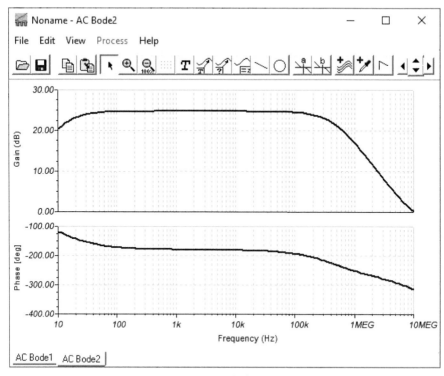

Figure 2.174

Let's measure the bandwidth of the amplifier. We need to find the frequencies that mid-band gain decreased by −3 dB. According to Figure 2.175 and Figure 2.176, these frequencies are 13.17 Hz and 442.61 kHz. So, the bandwidth is $442.61 \text{ kHz} - 13.17 \text{ Hz} - 442.61 \text{ kHz}$. Note that the bandwidth of the amplifier decreases considerably in comparison to Example 20.

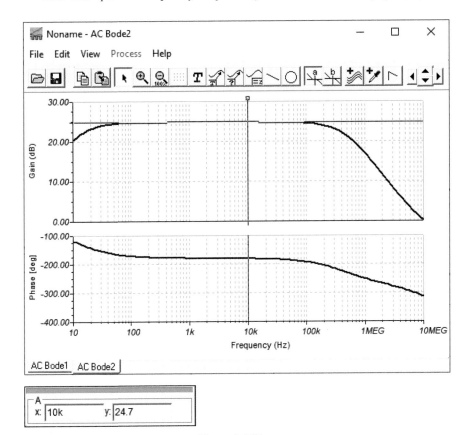

Figure 2.175

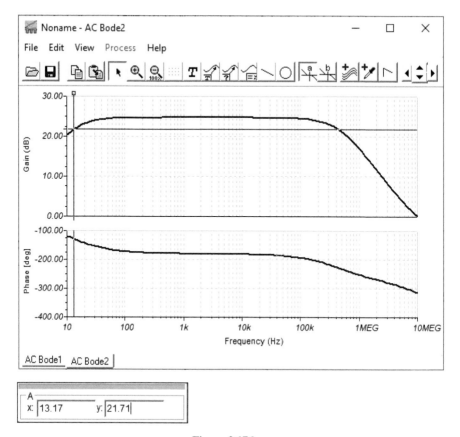

Figure 2.176

2.23 Example 22: Input Impedance of Common Emitter Amplifier

We learned how to draw the graph of the input impedance of an electric circuit with the aid of MATLAB in Example 29 of Chapter 1. In this example, we want to draw the graph of the input impedance of common emitter amplifier of Example 16 on the [10 Hz, 10 MHz] range. In this example, we will do all of the jobs in the TINA-TI environment.

Let's start. Open the schematic of Example 16 and change it to what is shown in the Figure 2.177. Note that $Z_{in}(j\omega) = \frac{V_{in}(j\omega)}{I_{in}(j\omega)}$. Therefore, we need to measure the input voltage (V_{in}) and input current (I_{in}) of the amplifier.

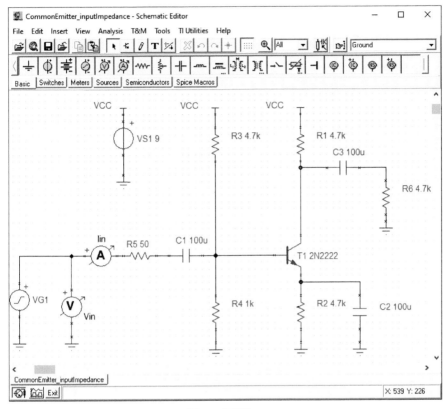

Figure 2.177

Click the Analysis > AC Analysis > AC Transfer Characteristic (Figure 2.178). This opens the AC Transfer Characteristic window (Figure 2.179). Run the simulation with the settings shown in Figure 2.179. The simulation result is shown in Figure 2.180.

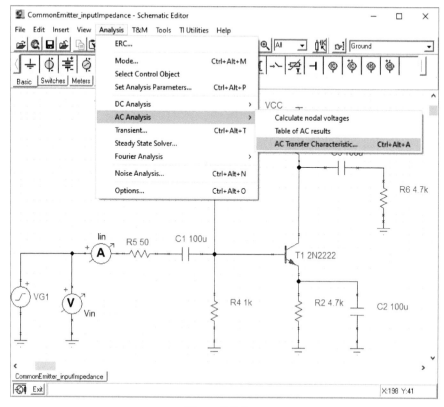

Figure 2.178

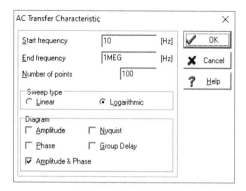

Figure 2.179

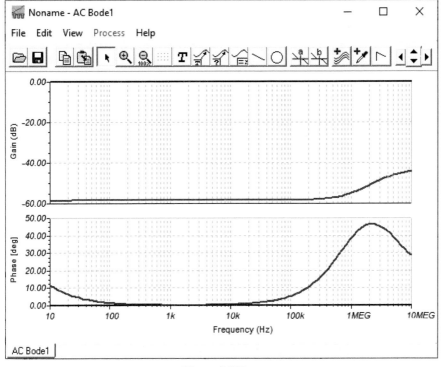

Figure 2.180

Click the Post-processor icon (Figure 2.181). This opens the post-processor window (Figure 2.182).

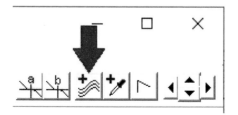

Figure 2.181

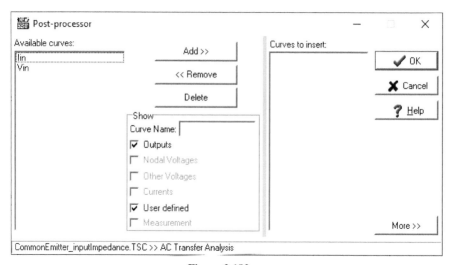

Figure 2.182

Click the More button in Figure 2.182. After clicking the More button, the Post-processor window changes to what is shown in Figure 2.183.

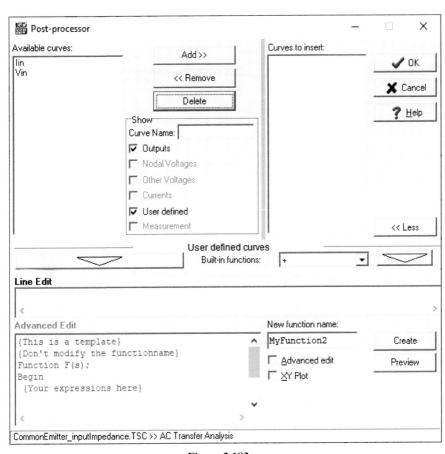

Figure 2.183

Click on the V_{in} to select it. Then click the down arrow button. This adds the 'Vin(s)' to the Line Edit box (Figure 2.184).

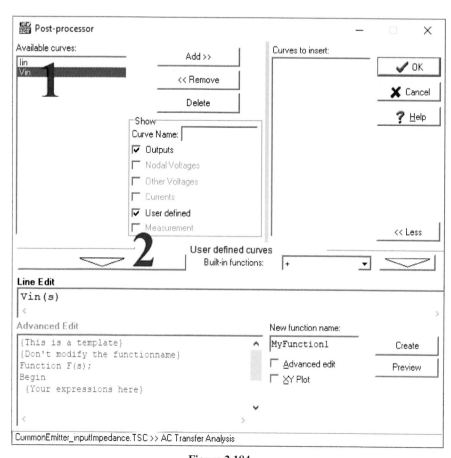

Figure 2.184

Add a '/' to the Line Edit box (Figure 2.185).

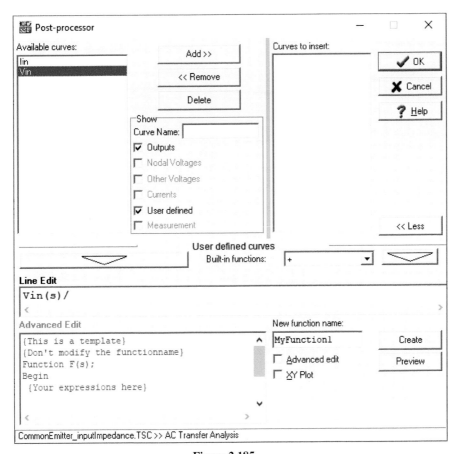

Figure 2.185

Click on the Iin to select it. Then click the down arrow button. This adds the 'Iin(s)' to the Line Edit box (Figure 2.186).

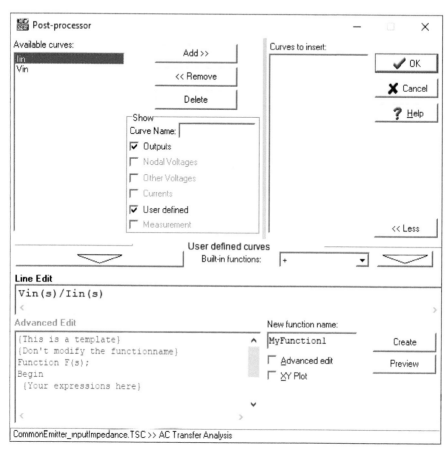

Figure 2.186

Enter the 'Zin' into the new function name box and click the Create button (Figure 2.187). This adds the 'Zin' to the Curves to insert box (Figure 2.188).

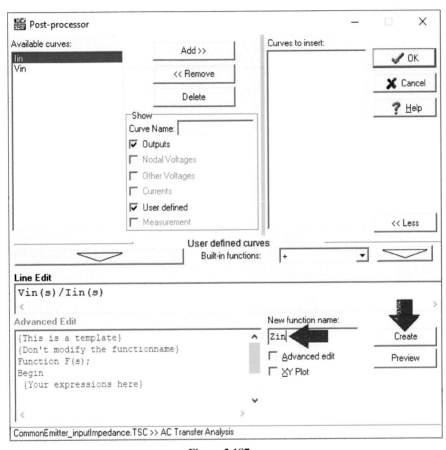

Figure 2.187

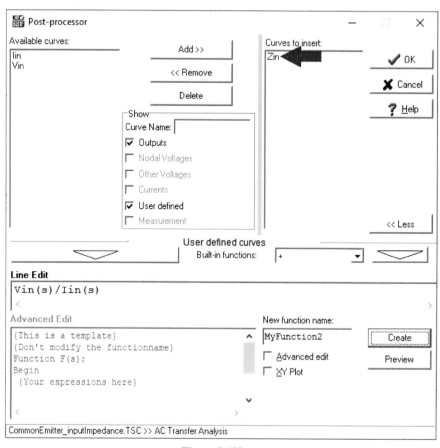

Figure 2.188

Click the OK button in Figure 2.188 to run the simulation. The simulation result is shown in Figure 2.189.

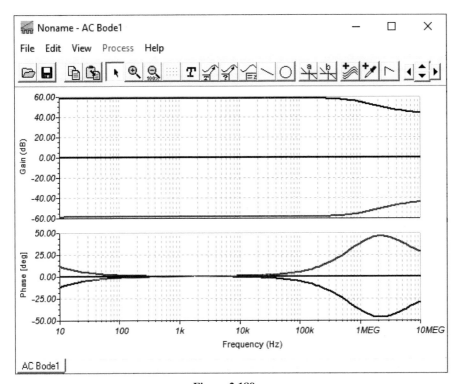

Figure 2.189

Click the View> Show/Hide curves (Figure 2.190). This opens the Show/Hide curves window. Uncheck the Vin and Iin boxes and click the Close button (Figure 2.191). Now, the only input impedance is shown on the screen (Figure 2.192).

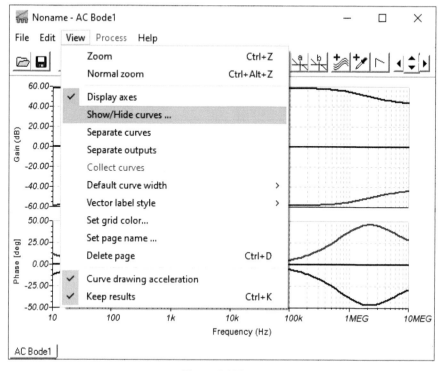

Figure 2.190

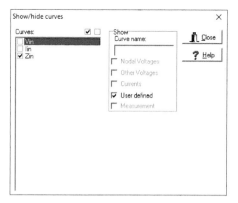

Figure 2.191

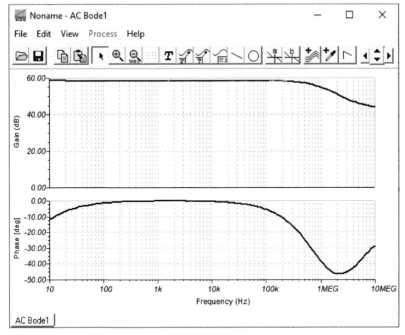

Figure 2.192

Note that the vertical axis in Figure 2.192 is in dB. Let's convert it into Ohm. In order to do this, double click on the vertical gain axis and change the Scale into 'Linear' (Figure 2.193).

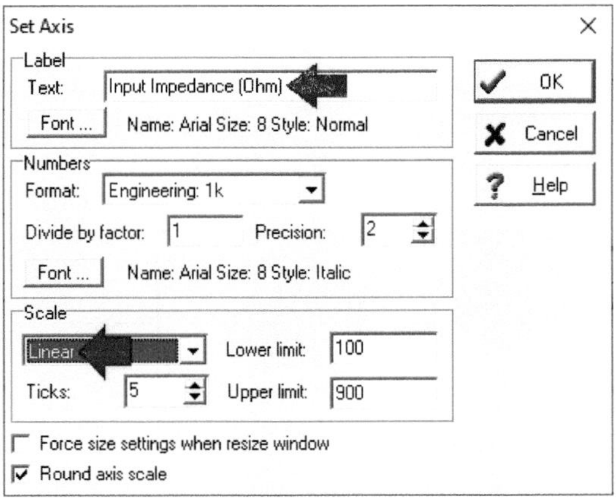

Figure 2.193

After clicking the OK button in Figure 2.193, the graph changes to what is shown in the Figure 2.194. The vertical axis of the gain (magnitude) graph is in Ohms.

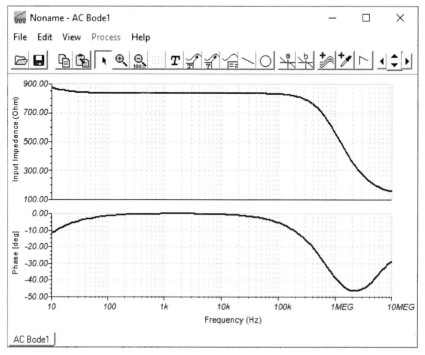

Figure 2.194

You can use the cursors to read the graph shown in Figure 2.195. For instance, at 1 kHz, the gain (magnitude) graph value is 835.83 Ω and the phase graph value is -195.87 m degrees. Therefore the input impedance at 1 kHz is $835.83e^{-j0.19587°} = 835.8251 - j2.8573\Omega$.

Figure 2.195

2.24 Example 23: Output Impedance of Common Emitter Amplifier

The output impedance of the common emitter amplifier of Example 16 can be found with the aid of the schematic shown in the Figure 2.196. Details of obtaining the graph of output impedance are similar to Example 22 and it is not explained here. Graph of output impedance is shown in the Figure 2.197.

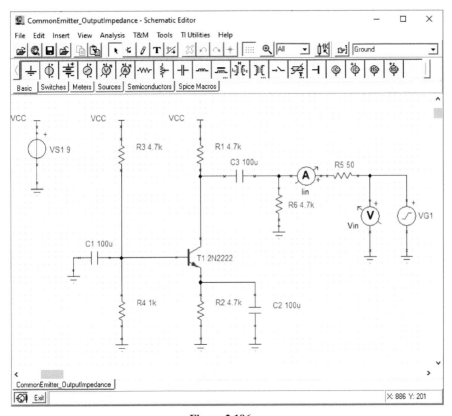

Figure 2.196

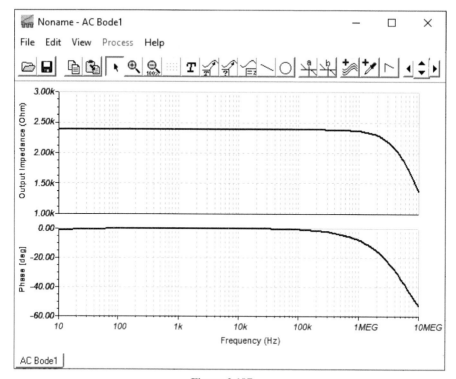

Figure 2.197

2.25 Example 24: Measurement of Input/Output Impedance with Ohm Meter Block

In Examples 22 and 23 we learned how to draw the graph of input/output impedance. Sometimes you don't need the whole graph but you need to know the magnitude of impedance at a specific frequency. In such cases, you can use the Ohm meter block to measure the magnitude of impedance at the frequency you want.

Let's study an example. Assume that we want to know the magnitude of input impedance at 1 kHz. The schematic shown in Figure 2.198 can be used for this purpose.

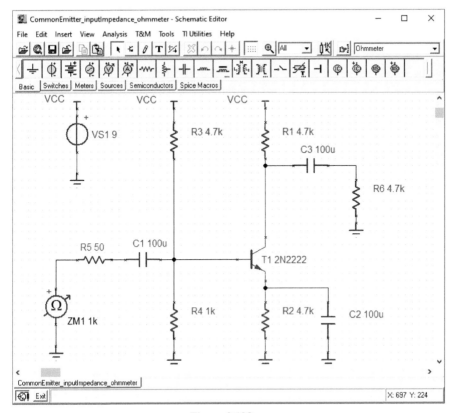

Figure 2.198

Click the Analysis> AC Analysis> Calculate nodal voltages (Figure 2.199). The simulation result is shown in Figure 2.200. According to Figure 2.200, the magnitude of input impedance at 1 kHz is 835.83 Ω. The obtained result is quite close to the number shown in Figure 2.195.

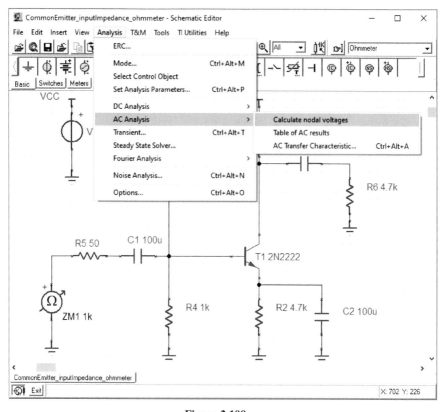

Figure 2.199

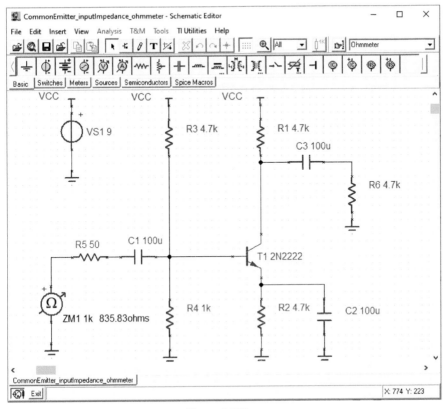

Figure 2.200

2.26 Example 25: Modeling a Custom Bipolar Transistor

You can make a custom transistor custom transistor in TINA-TI easily. In order to do so, double-click the bipolar transistor and click the three dots behind the Type (Figure 2.201). Then select the! NPN (Figure 2.202) and enter your custom values to the "model parameters" list.

For instance, the value of Early voltage must be entered into 'Forw. early voltage' box (Figure 2.202) and value of current gain must be entered into the 'Forward beta' box (Figure 2.203).

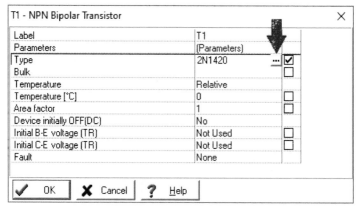

Figure 2.201

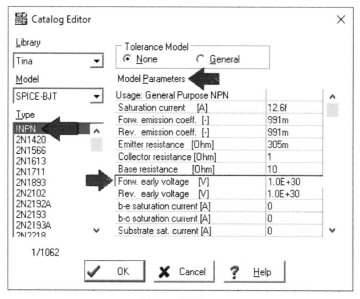

Figure 2.202

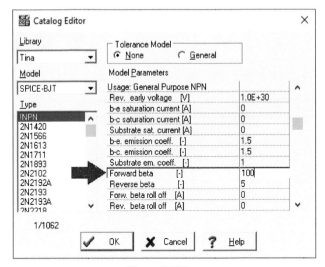

Figure 2.203

If you want to model a custom PNP transistor, you need to add a PNP transistor (Figure 2.204) to the schematic and select! PNP. Other steps are similar to the NPN case.

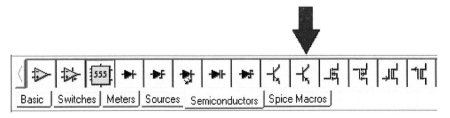

Figure 2.204

2.27 Example 26: Modeling a Custom Field Effect Transistor

Modeling a custom field-effect transistor is quite easy in TINA-TI. Just add the transistor (Figure 2.205) to the schematic, double click on it, click the three dots behind the Type box and select the 'No name' (Figure 2.206). Now you can enter your values into the 'Model Parameters' list.

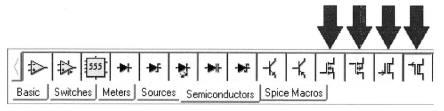

Figure 2.205

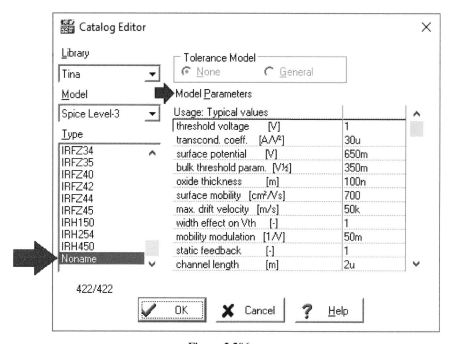

Figure 2.206

2.28 Example 27: Generating the List of Circuit Components

TINA-TI can extract the list of components used in the schematic. This is very useful when you want to make the circuit. In this example, we want to see the list of components of Example 16. In order to do this, open the schematic of Example 16 and click the File > Bill of Materials (Figure 2.207). The result is shown in Figure 2.208. You can click the Print button in order to print the list or you can click the Save button to save the list.

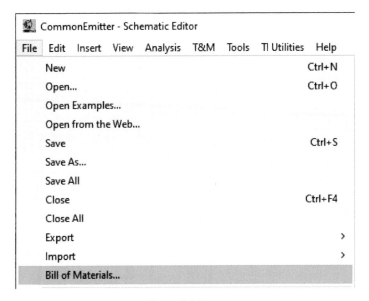

Figure 2.207

#	Quantity	Label	Value	Footprint	Parameter #1	Parameter #2	Parameter #3	Parameter #4
1	1	C1	100u	CP_CYL300_D700_L1400				
2	1	C2	100u	CP_CYL300_D700_L1400				
3	1	C3	100u	CP_CYL300_D700_L1400				
4	1	R4	1k	R_AX600_W200				
5	1	T1	2N2222	TO206AA				
6	1	R1	4.7k	R_AX600_W200				
7	1	R2	4.7k	R_AX600_W200				
8	1	R3	4.7k	R_AX600_W200				
9	1	R6	4.7k	R_AX600_W200				
10	1	R5	50	R_AX600_W200				
11	1	VS1	9	Voltgen				
12	1	VG1		Sgen				

Figure 2.208

2.29 Example 28: Non Inverting op amp Amplifier

In this example we want to simulate the circuit shown in Figure 2.209. The gain of this amplifier is $1 + \frac{R_1}{R_2} = 1 + 9 = 10$. So, we expect to observe 100 mV at the output of the amplifier.

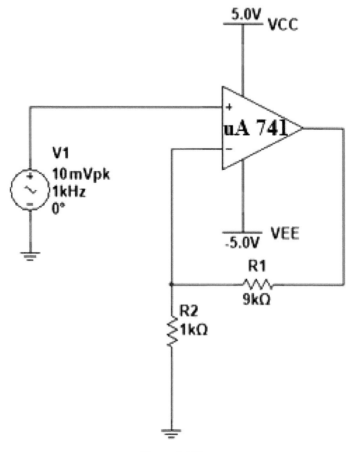

Figure 2.209

Let's start. Draw the schematic shown in the Figure 2.210. The required op amp block can be found in the 'Semiconductors' and 'Spice Macros' tabs (Figure 2.211 and Figure 2.212). 'VCC' and 'VEE' are Jumper blocks (Figure 2.213).

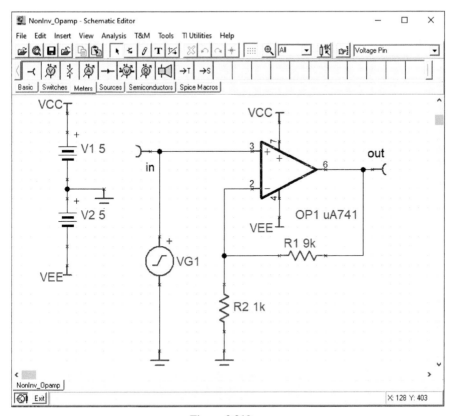

Figure 2.210

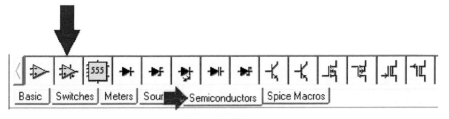

Figure 2.211

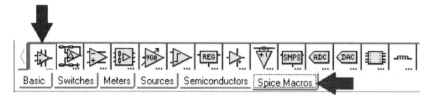

Figure 2.212

Figure 2.213

Click the Analysis > Transient (Figure 2.214) and run the simulation with the settings shown in Figure 2.215. The result is shown in Figure 2.216. Note that the output waveform is not symmetric. It has a DC offset.

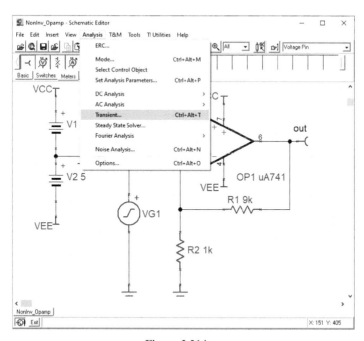

Figure 2.214

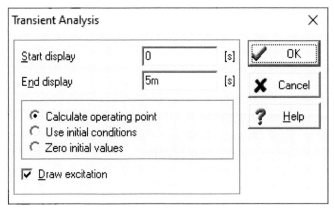

Figure 2.215

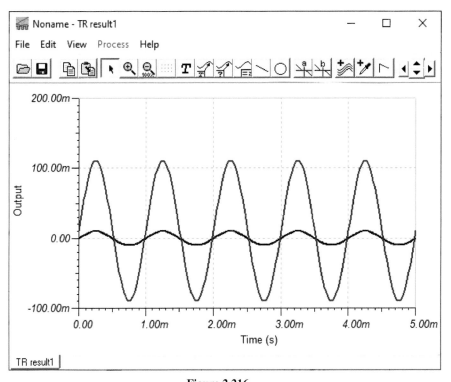

Figure 2.216

Let's measure the value of DC offset. Remove VG1 and connect the input to the ground (Figure 2.217). Run the transient analysis simulation with settings shown in Figure 2.215. The simulation result is shown in the Figure 2.218. According to Figure 2.218, the offset is about 10 mV.

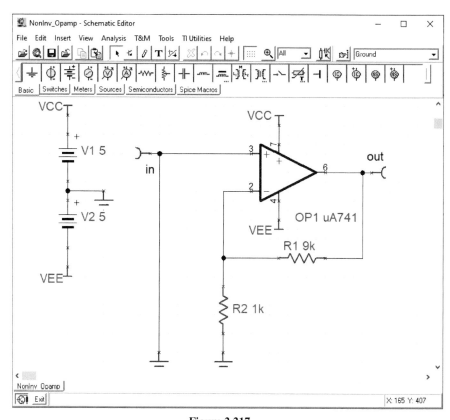

Figure 2.217

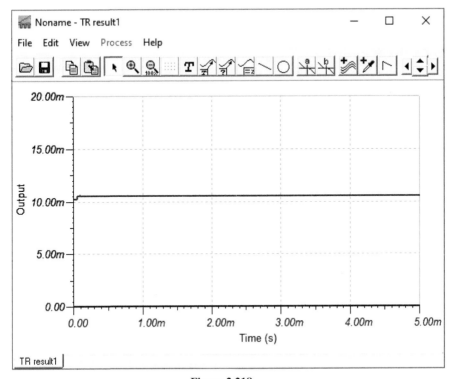

Figure 2.218

Let's find a way to get rid of this offset. Add a 10 kΩ resistor in series with the positive terminal of op amp (Figure 2.219) and rerun the transient simulation with the settings shown in Figure 2.215. The simulation result is shown in Figure 2.220. The DC offset decreased to about 2.5 mV.

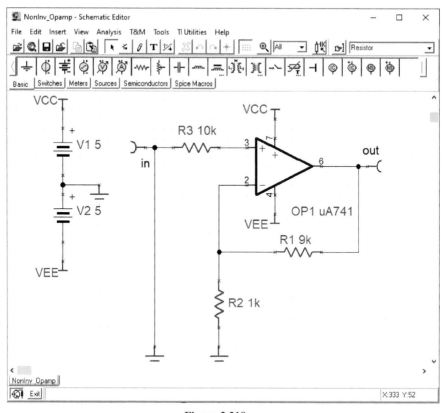

Figure 2.219

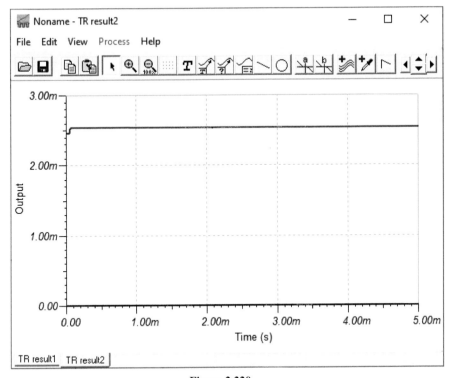

Figure 2.220

If you increase the series resistor to 13.2 kΩ (Figure 2.221), DC offset of the output decreases to −20 μV (Figure 2.222).

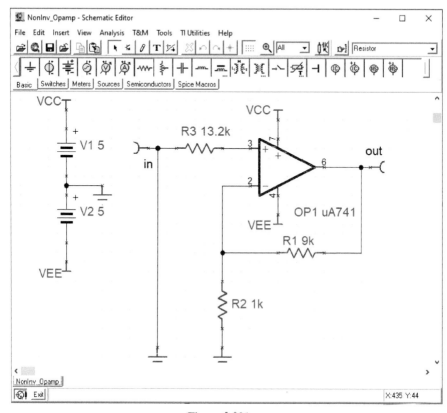

Figure 2.221

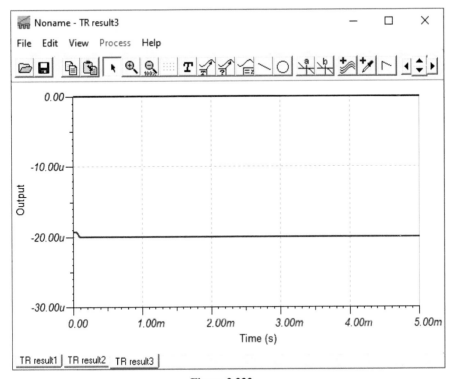

Figure 2.222

Now add the Voltage Generator VG1 to the schematic (Figure 2.223) and rerun the transient analysis with the settings shown in Figure 2.225. The simulation result is shown in Figure 2.224. Note that the output is symmetric now. The peak value of the output is 100 mV. So, the gain of an amplifier is 10 as expected.

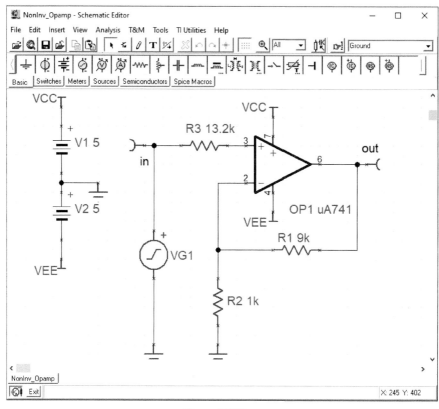

Figure 2.223

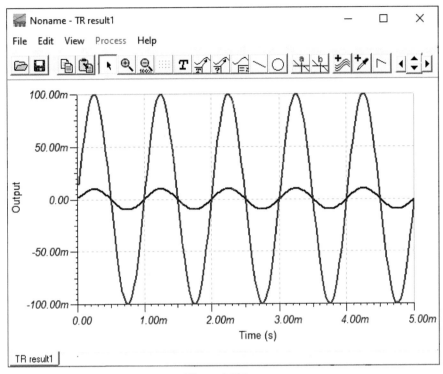

Figure 2.224

2.30 Example 29: Stability of op amp Amplifiers

In this example, we want to study the stability of the non-inverting op amp amplifier stability of non-inverting op amp amplifier of Example 28. We want to measure the phase margin of the circuit. Open the connection between the negative terminal of op amp and the feedback network and put a Voltage Generator block there (Figure 2.225). Note that the input of the amplifier must be zero in studying the stability. That is why the + terminal of the op amp is grounded.

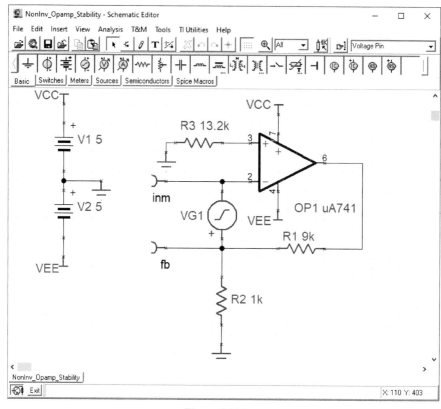

Figure 2.225

Click the Analysis > AC Analysis > AC Transfer Characteristic (Figure 2.226) and run the simulation with the settings shown in Figure 2.227. The simulation result is shown in Figure 2.228.

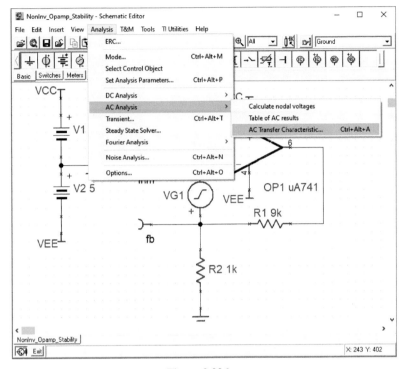

Figure 2.226

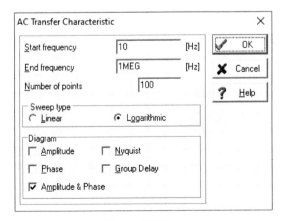

Figure 2.227

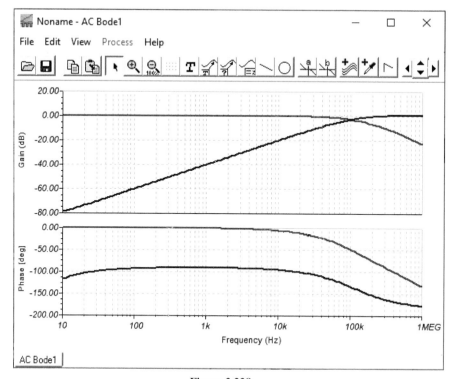

Figure 2.228

Click the Post-processor icon (Figure 2.229) and do the settings similar to Figure 2.230.

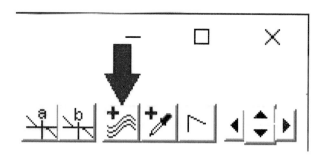

Figure 2.229

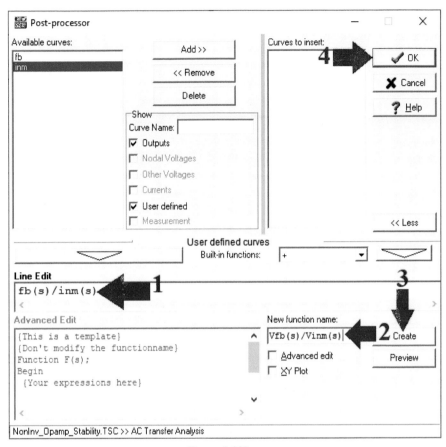

Figure 2.230

After clicking the OK button in Figure 2.230, the result shown in Figure 2.231 appears on the screen.

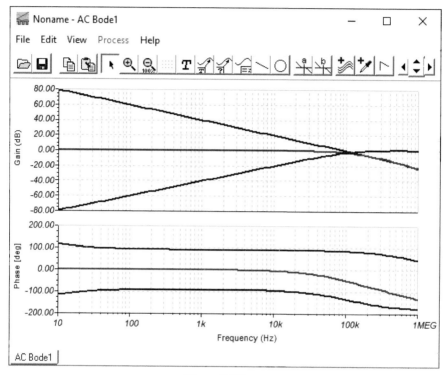

Figure 2.231

Click the View > Show/Hide curves and uncheck the 'inm' and 'fb' boxes (Figure 2.232). After clicking the Close button, the frequency response of $\frac{fb(j\omega)}{inm(j\omega)}$ appears on the screen (Figure 2.233).

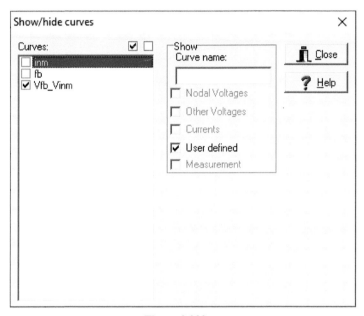

Figure 2.232

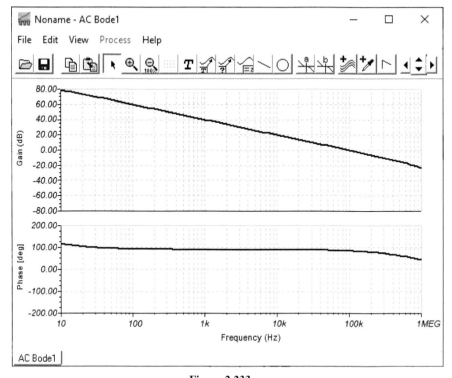

Figure 2.233

Right-click on the frequency response graph and click the '0 dB Cross-over Frequency' (Figure 2.234). This calculates the intersection of the graph with a 0 dB line. According to Figure 2.235, the intersection occurred at 98.21 kHz.

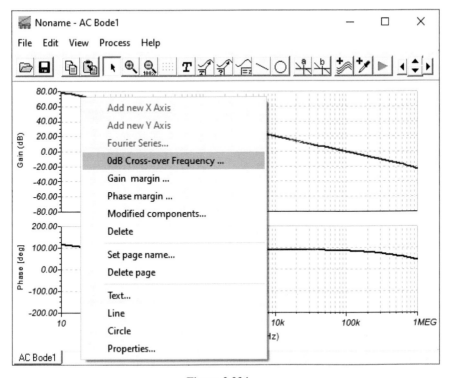

Figure 2.234

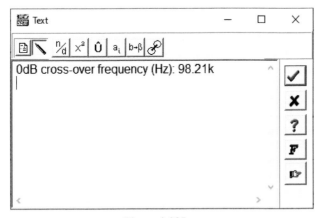

Figure 2.235

Right-click on the frequency response graph and click the 'Phase margin' (Figure 2.236). According to Figure 2.237, the phase margin is 84.42°. So, the stability of this amplifier is acceptable. Generally, a phase margin of bigger than 45° and a gain margin of bigger than 6 dB is required.

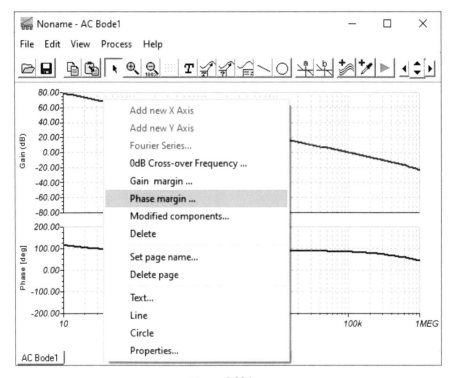

Figure 2.236

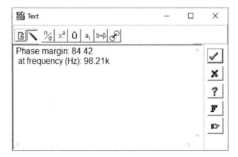

Figure 2.237

Right-click on the frequency response graph and click the 'Gain margin' (Figure 2.238). After clicking, the error message shown in Figure 2.239 appears on the screen. Note that the phase graph does not reach $-180°$. Therefore, the gain margin of this loop is infinity. That is why we received the error message shown in Figure 2.239.

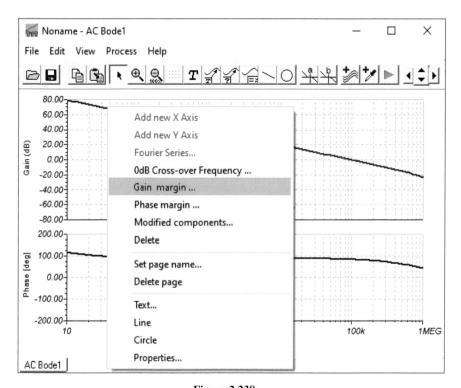

Figure 2.238

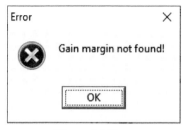

Figure 2.239

2.31 Example 30: Measurement of DC Operating Point of a Differential Pair Amplifier

In this example, we want to measure the operating point of the differential pair CMRR of differential pair amplifier shown in Figure 2.240.

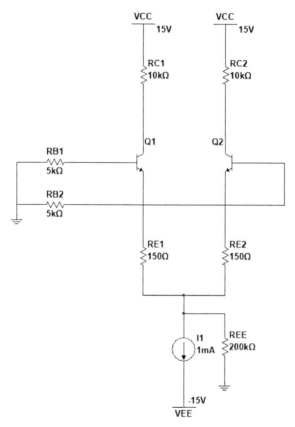

Figure 2.240

Let's start. Draw the schematic shown in Figure 2.241 and click the Analysis > DC Analysis > Table of DC results (Figure 2.242). The result of the simulation is shown in Figure 2.243.

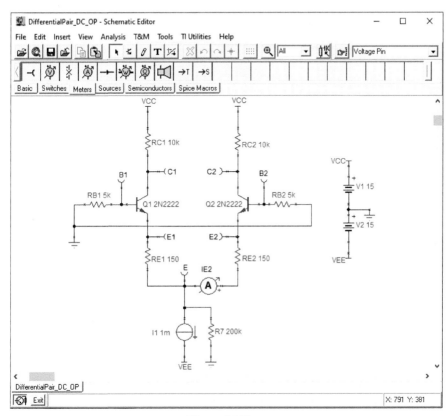

Figure 2.241

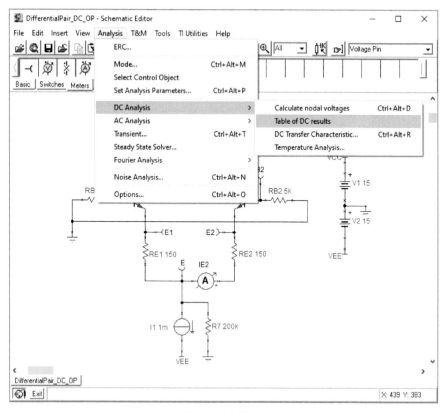

Figure 2.242

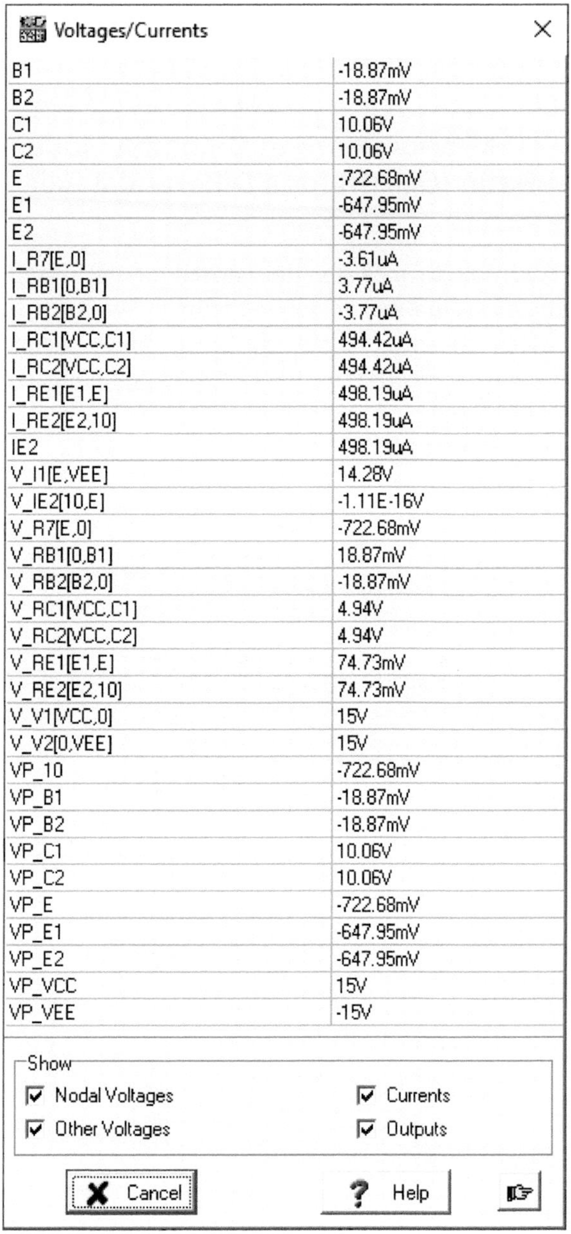

Voltages/Currents	✕
B1	-18.87mV
B2	-18.87mV
C1	10.06V
C2	10.06V
E	-722.68mV
E1	-647.95mV
E2	-647.95mV
I_R7[E,0]	-3.61uA
I_RB1[0,B1]	3.77uA
I_RB2[B2,0]	-3.77uA
I_RC1[VCC,C1]	494.42uA
I_RC2[VCC,C2]	494.42uA
I_RE1[E1,E]	498.19uA
I_RE2[E2,10]	498.19uA
IE2	498.19uA
V_I1[E,VEE]	14.28V
V_IE2[10,E]	-1.11E-16V
V_R7[E,0]	-722.68mV
V_RB1[0,B1]	18.87mV
V_RB2[B2,0]	-18.87mV
V_RC1[VCC,C1]	4.94V
V_RC2[VCC,C2]	4.94V
V_RE1[E1,E]	74.73mV
V_RE2[E2,10]	74.73mV
V_V1[VCC,0]	15V
V_V2[0,VEE]	15V
VP_10	-722.68mV
VP_B1	-18.87mV
VP_B2	-18.87mV
VP_C1	10.06V
VP_C2	10.06V
VP_E	-722.68mV
VP_E1	-647.95mV
VP_E2	-647.95mV
VP_VCC	15V
VP_VEE	-15V

Show
☑ Nodal Voltages   ☑ Currents
☑ Other Voltages   ☑ Outputs

✕ Cancel   ? Help   ☞

Figure 2.243

According to Figure 2.243, operating point of transistor Q2 (or Q1) is (10.06- (−0.64795) V, 0.49819 mA) = (10.708 V, 0.49819 mA).

2.32 Example 31: Measurement of Common Mode Rejection Ratio (CMRR) for a Differential Pair Amplifier

In this example, we want to measure the CMRR of the differential pair CMRR of differential pair amplifier shown in Figure 2.244. The two collector resistances are accurate to within $\pm1\%$. So, their values change between $10 \times 0.99 = 9.9k\Omega$ and $10 \times 1.01 = 10.1k\Omega$. Other components (transistor Q1 and Q2, resistor RB1 and Rb2, resistor RE1 and RE2) are assumed to be identical.

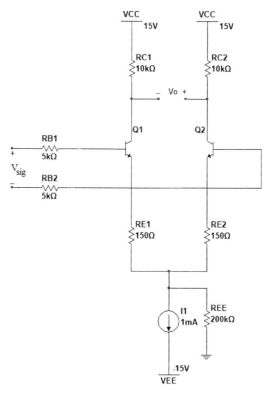

Figure 2.244

Let's measure the common-mode gain of the amplifier. In order to do this, add the voltage source VG1 to the schematic (Figure 2.245). Settings of the voltage source VG1 are shown in Figure 2.246. Note that in Figure 2.245, RC1=9.9 kΩ and RC2=10.1 kΩ. If you run the schematic of Figure 2.245, with RC1=RC2=10 kΩ, the output (voltage difference between the collectors) becomes zero since the circuit is symmetric and the input is the same for both transistors. When the output for common-mode signal is zero, the common-mode gain is zero and the CMRR becomes infinity which is desired. However, obtaining a completely symmetric circuit is very difficult (if not impossible) in the real world.

We set the RC1 to 9.9 kΩ and RC2 to 10.1 kΩ in order to study the worst case (the biggest common-mode gain is obtainable for these values).

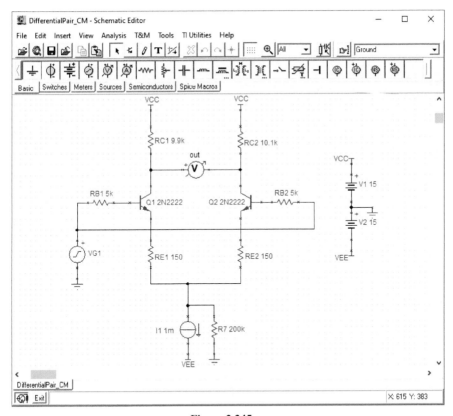

Figure 2.245

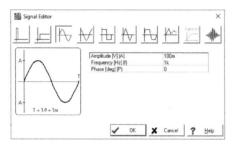

Figure 2.246

Click the Analysis > Transient (Figure 2.247) and run a transient analysis with the settings shown in Figure 2.248. The simulation result is shown in Figure 2.249.

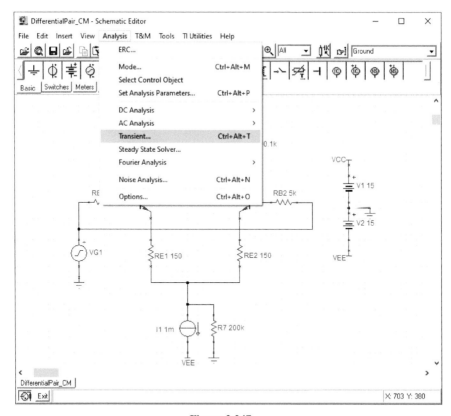

Figure 2.247

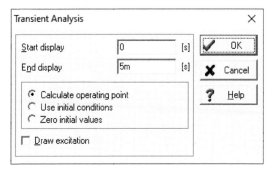

Figure 2.248

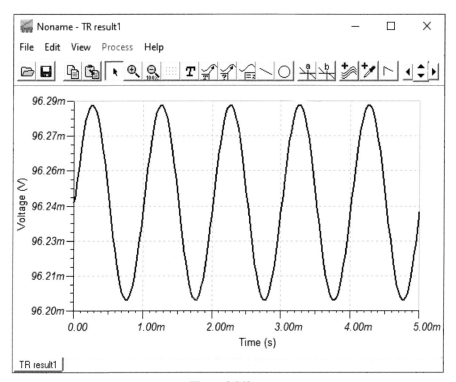

Figure 2.249

Let's measure the peak of output voltage. According to Figure 2.250, peak of the output voltage is 83.43 μV.

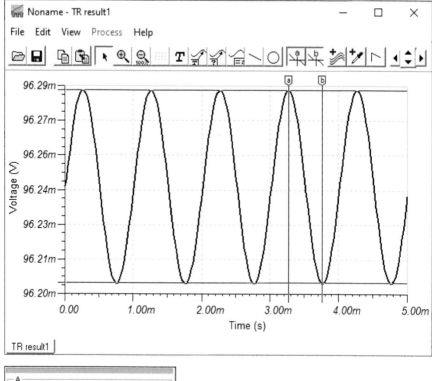

Figure 2.250

The common-mode gain is the ratio of peak of the output voltage to peak of the input voltage. Peak peak of the input voltage is $2 \times 100 \ mV = 200 \ mV$. According to Figure 2.251, common-mode gain is 4.1715×10^{-4}.

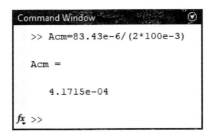

Figure 2.251

It is time to measure the differential mode gain. Change the schematic to what is shown in Figure 2.252. Settings of V1 and V2 are shown in Figure 2.253 and Figure 2.254, respectively. Note that 180° of phase difference exists between these two voltage sources.

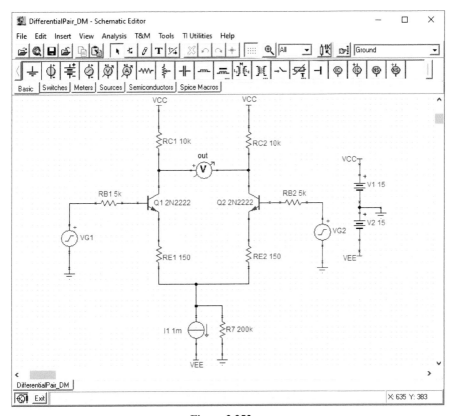

Figure 2.252

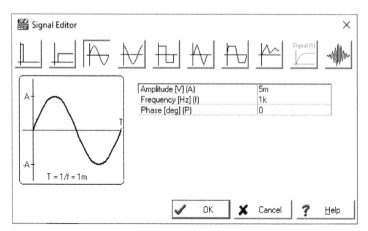

Figure 2.253

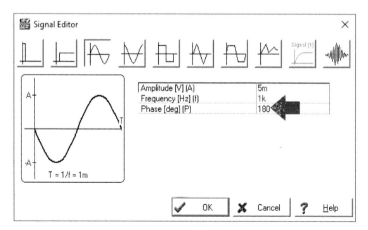

Figure 2.254

Run a transient analysis with the settings shown in Figure 2.255. The simulation result is shown in Figure 2.256.

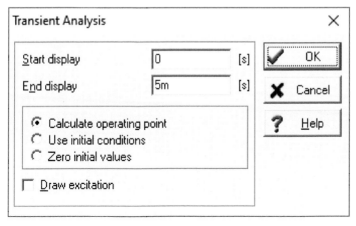

Figure 2.255

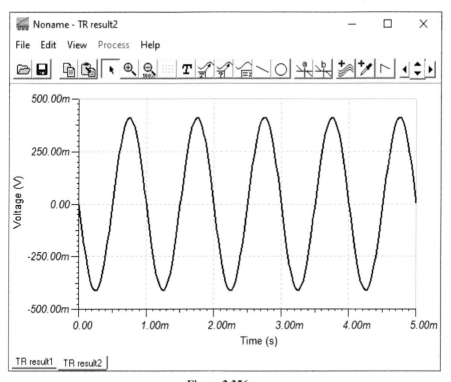

Figure 2.256

Let's measure the peak of the output voltage. According to Figure 2.257, peak of the output voltage is 819.36 mV.

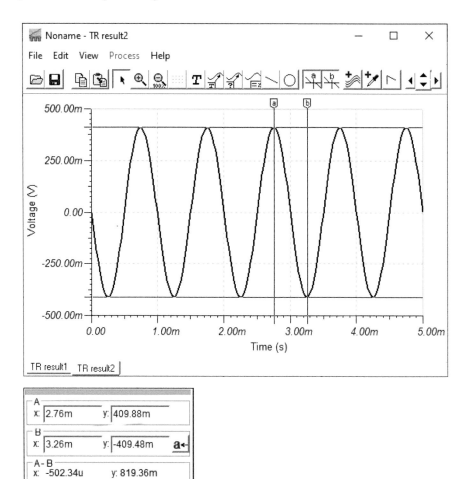

Figure 2.257

The schematic is shown in the Figure 2.258 measures the peak of input voltage. Run a transient analysis with the settings shown in the Figure 2.259. The simulation result is shown in the Figure 2.260. According to Figure 2.260, peak of the output is 20 mV.

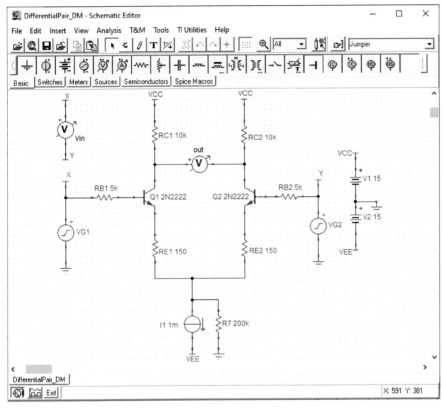

Figure 2.258

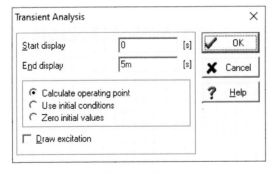

Figure 2.259

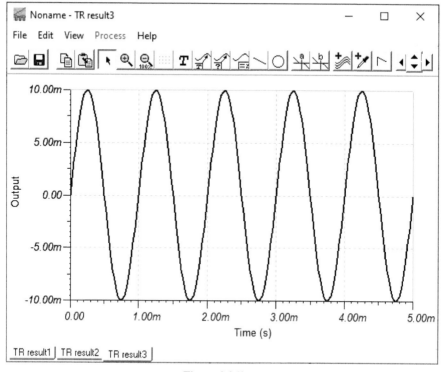

Figure 2.260

The differential mode gain is the ratio of peak of the output voltage to peak of the input voltage. According to Figure 2.261, differential mode gain is 40.9680.

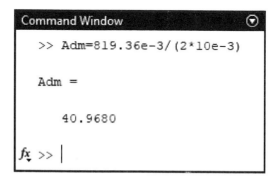

Figure 2.261

According to Figure 2.262, the worst case (minimum) CMRR is 99.8431 dB. Note that we used the maximum common-mode gain in the calculation of the CMRR. So, minimum CMRR is obtained. When the difference between the RC1 and RC2 decreases, the CMRR becomes bigger than 99.8431dB. For instance, you can simulate the circuit for RC1=9.99 kΩ and RC2=10.01 kΩ and observe the increase in CMRR.

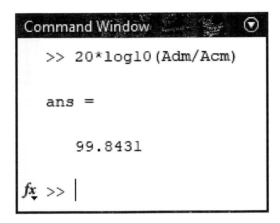

Figure 2.262

2.33 Example 32: Astable Oscillator

In this example we want to simulate an astable oscillator with 555 IC (Figure 2.263).

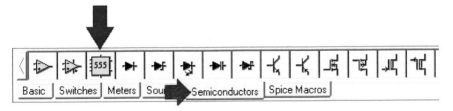

Figure 2.263

Draw the schematic shown in Figure 2.264.

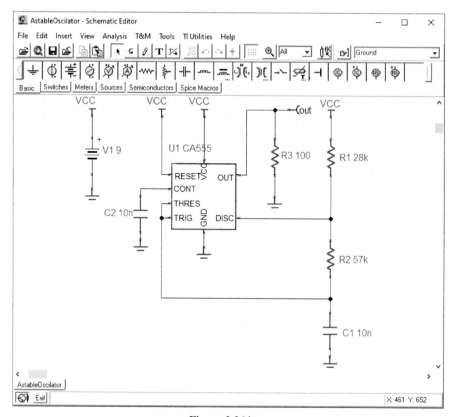

Figure 2.264

Run a transient analysis with the settings shown in Figure 2.265. The simulation result is shown in Figure 2.266.

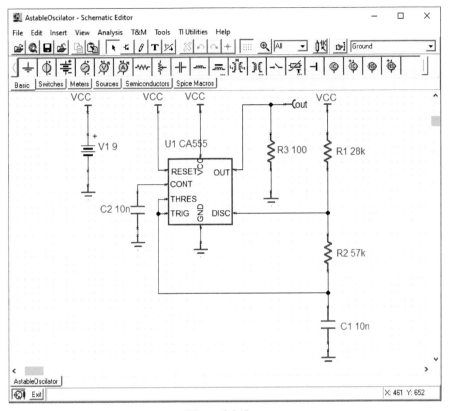

Figure 2.265

Figure 2.266

Let's measure the width of a low and high portions of the waveform. According to Figure 2.267, the width of a low portion is 393.78 μs. According to Figure 2.268, the width of a high portion is 590.81 μs. So, the frequency of the waveform is $\frac{1}{393.87\ \mu s + 590.81\ \mu s} = 1.016$ kHz.

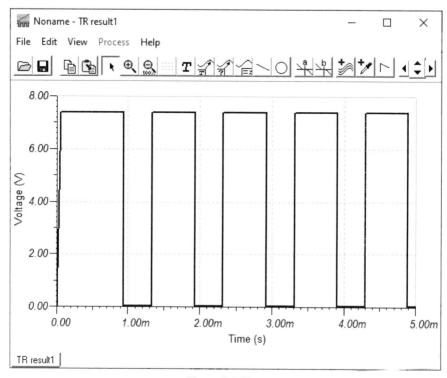

Figure 2.267

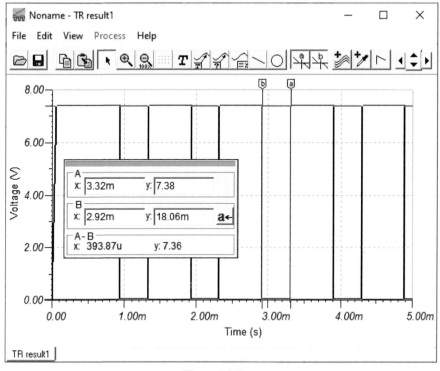

Figure 2.268

Let's check the obtained results. The calculation shown in Figure 2.269 shows that TINA-TI results are correct.

```
Command Window -                                              ⊙
    >> R1=28e3;R2=57e3;C=10e-9;
    >> HighDuration_us=0.693*(R1+R2)*C/1e-6

    HighDuration_us =

        589.0500

    >> LowDuration_us=0.693*R2*C/1e-6

    LowDuration_us =

        395.0100

    >> frequency=1/((HighDuration_us+LowDuration_us)*1e-6)

    frequency =

        1.0162e+03

fx >> |
```

Figure 2.269

2.34 Example 33: Buck Converter

In this example, we want to simulate a buck converter. Draw the schematic shown in Figure 2.270. The MOSFET and the diode can be added to the schematic with the aid of icons shown in Figure 2.271.

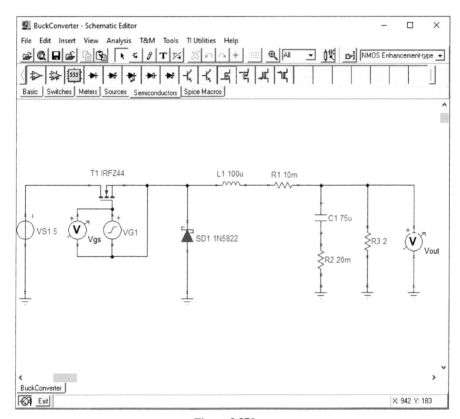

Figure 2.270

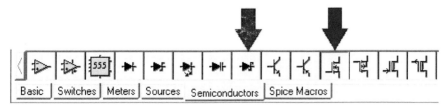

Figure 2.271

Settings of Voltage Generator VG1 are shown in Figure 2.272 and Figure 2.273. These settings generate the waveform shown in Figure 2.274. Note that VG1 controls the MOSFET. According to Figure 2.274, the switching frequency is $\frac{1}{40\mu} = 25$ kHz and duty cycle of the pulse applied to the MOSFET is $\frac{20\mu}{40\mu} = 0.5$.

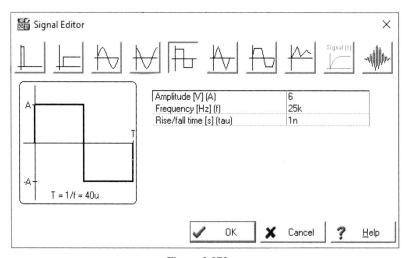

Figure 2.272

Figure 2.273

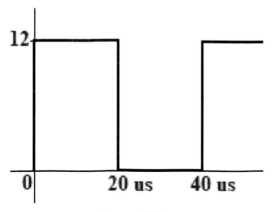

Figure 2.274

Run a transient analysis with the settings shown in Figure 2.275. The simulation result is shown in Figure 2.276.

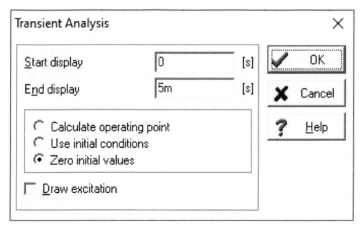

Figure 2.275

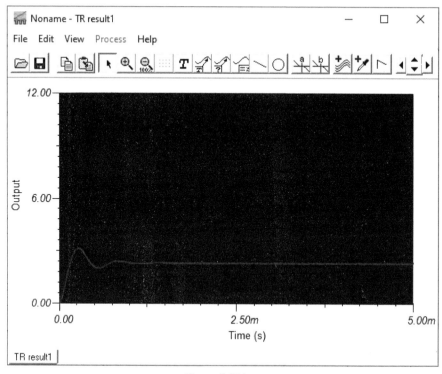

Figure 2.276

Click the View > Show/Hide curves (Figure 2.277). Then uncheck the V_{out} box (Figure 2.278). This hides the waveform of the output voltage (Figure 2.279).

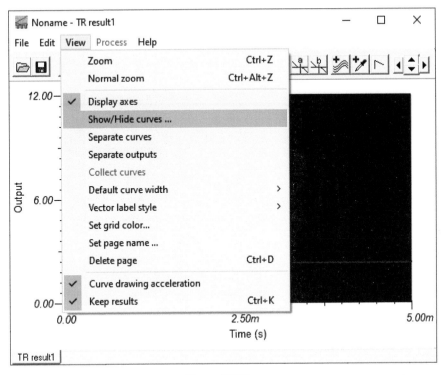

Figure 2.277

Figure 2.278

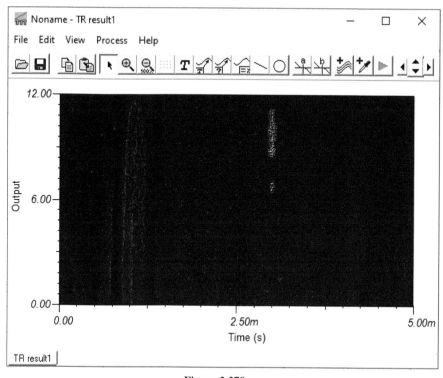

Figure 2.279

Use the magnifier icon to see the gate-source voltage better (Figure 2.280). According to Figure 2.281, the duty cycle of the pulse is around 50% as expected.

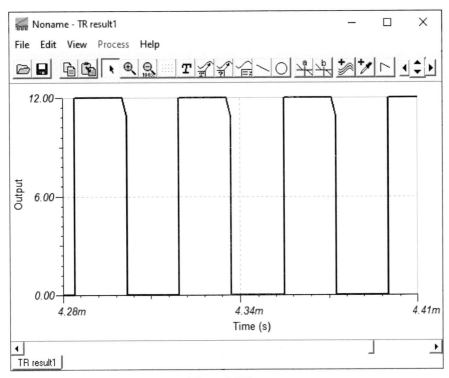

Figure 2.280

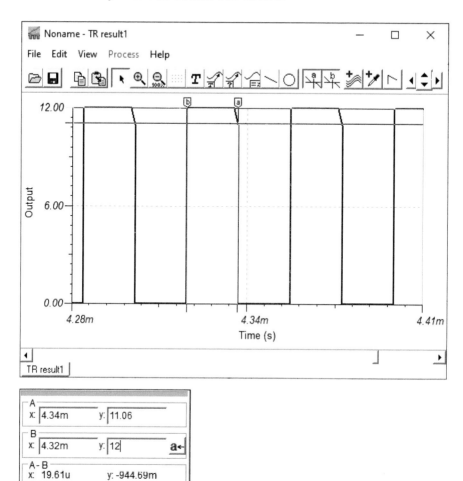

Figure 2.281

Click the View > Show/Hide curves (Figure 2.282). Then uncheck the Vgs box (Figure 2.283). This hides the waveform of gate-source (Figure 2.284).

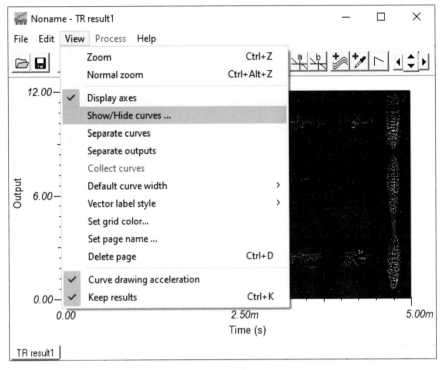

Figure 2.282

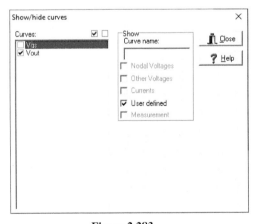

Figure 2.283

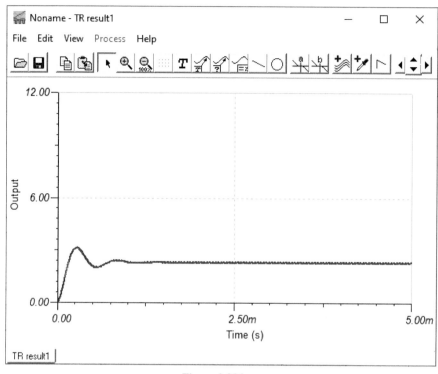

Figure 2.284

Click the Normal zoom icon to obtain a better view (Figure 2.285).

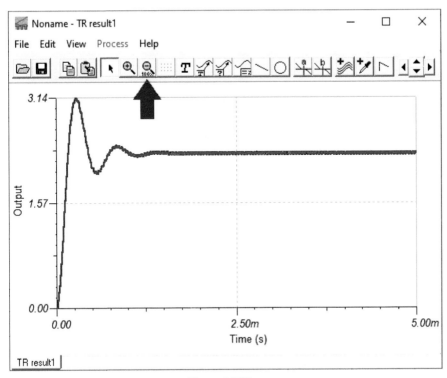

Figure 2.285

Let's measure the output voltage ripple. Use the Zoom icon (Figure 2.286) to zoom into the steady-state portion of the waveform (Figure 2.287).

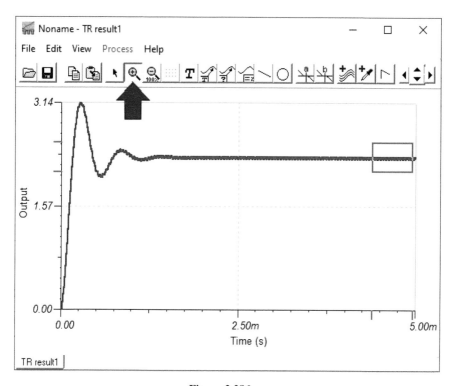

Figure 2.286

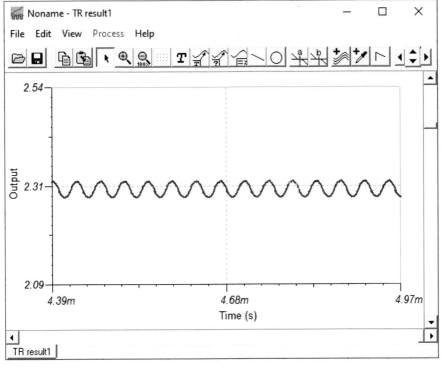

Figure 2.287

Use cursors to measure the peak of the output voltage ripple. According to Figure 2.288, peak of the output voltage ripple is 35.91 mV.

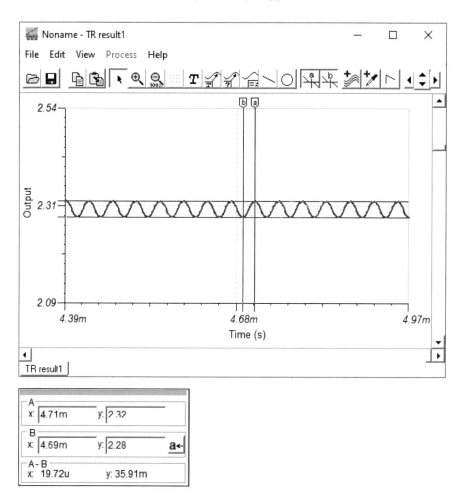

Figure 2.288

The average value of the output voltage can be found by averaging the minimum and maximum of the output voltage. According to Figure 2.288, the minimum and maximum output voltages are 2.28 V and 2.32 V, respectively. Therefore, an average value of the output voltage is $\frac{2.28+2.32}{2} = 2.30\ V$. You can use the Process > Averages to measure the average value of the output voltage as well (Figure 2.289).

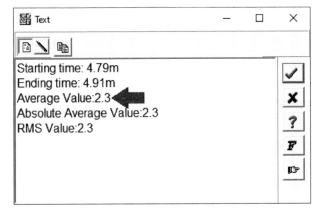

Figure 2.289

2.35 Example 34: Operating Mode of Converter

In this example, we want to determine the operating mode (Continu-ous/Discontinuous Conduction Mode) of the Buck converter in Example 33. We need to study the inductor current waveform in order to determine the operating mode of the converter. Therefore, change the schematic to what is shown in Figure 2.290.

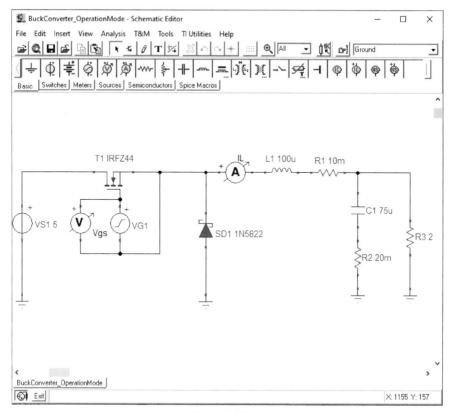

Figure 2.290

Run a transient analysis with the parameters shown in Figure 2.291. The simulation result is shown in Figure 2.292.

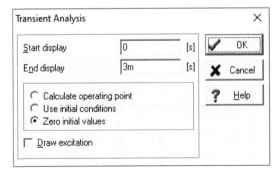

Figure 2.291

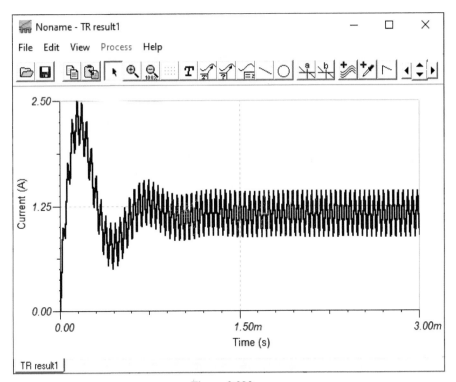

Figure 2.292

Use the magnifier icon to zoom into the steady-state portion of the waveform (Figure 2.293). According to Figure 2.293, the minimum of the waveform doesn't reach zero. Therefore, the converter is operated in Continuous Conduction Mode (CCM).

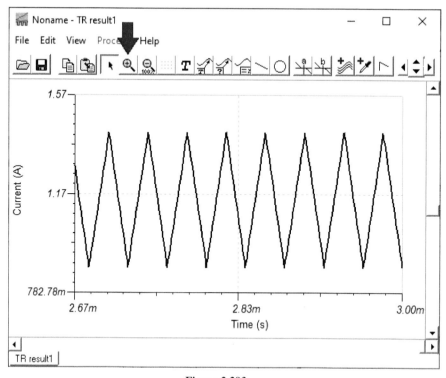

Figure 2.293

Close the window shown in Figure 2.293 and increase the output load to 25 Ω. Run the transient simulation with the settings shown in Figure 2.291. The simulation result is shown in Figure 2.294.

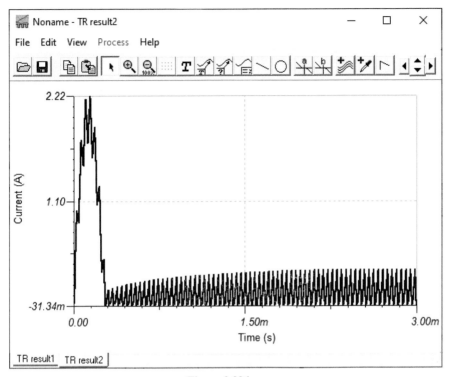

Figure 2.294

Zoom into the steady-state portion of the waveform (Figure 2.295). As shown in Figure 2.295, the minimum of the inductor current becomes negative. Therefore, the converter is operated in Discontinuous Conduction Mode (DCM).

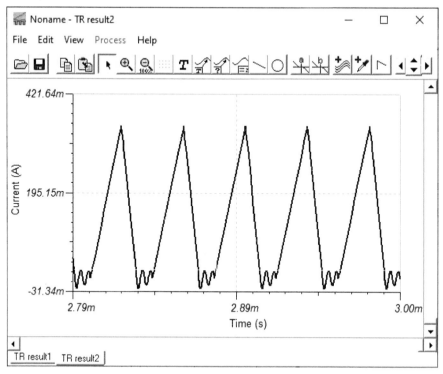

Figure 2.295

2.36 Example 35: Generating a Pulse with Desired Duty Cycle

We simulated a Buck converter in Example 33. In that example, we used a square wave Voltage Generator block to control the MOSFET. The square wave Voltage Generator block generates the duty cycle of 50% only. In this example, we will introduce a technique to generate a pulse with desired duty cycle.

Let's start. Start the MATLAB Editor (Figure 2.296).

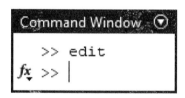

Figure 2.296

Type the following code in MATLAB Editor (Figure 2.297).

```
delete 'D:\gate_source_signal.txt'
f=1000*input('frequency in kHz?');
D=input('Duty cycle(0..1)?');
N=input('How many cycles?');
V=input('Output voltage?');
T=1/f;

time=[0 D*T 1.01*D*T T];
voltage=[V V 0 0];

fid=fopen('D:\gate_source_signal.txt','w');

for n=0:N
    s1=num2str(0+n*T);
    s2=' ';
    s3=num2str(V);
    fprintf(fid,[s1 s2 s3 '\n']);

    s1=num2str(D*T+n*T);
    s2=' ';
    s3=num2str(V);
    fprintf(fid,[s1 s2 s3 '\n']);

    s1=num2str(1.01*D*T+n*T);
    s2=' ';
    s3=num2str(0);
    fprintf(fid,[s1 s2 s3 '\n']);

    s1=num2str(T+n*T);
    s2=' ';
    s3=num2str(0);
    fprintf(fid,[s1 s2 s3 '\n']);
end
fclose('all');
```

```
 Editor - Untitled*                                                    ⊙ ×
  Untitled*  ×   +
  1        delete 'D:\gate_source_signal.txt'
  2        f=1000*input('frequency in kHz?');
  3        D=input('Duty cycle(0..1)?');
  4        N=input('How many cycles?');
  5        V=input('Output voltage?');
  6        T=1/f;
  7
  8        time=[0 D*T 1.01*D*T T];
  9        voltage=[V V 0 0];
 10
 11        fid=fopen('D:\gate_source_signal.txt','w');
 12
 13  ☐ for n=0:N
 14          s1=num2str(0+n*T);
 15          s2=' ';
 16          s3=num2str(V);
 17          fprintf(fid,[s1 s2 s3 '\n']);
 18
 19          s1=num2str(D*T+n*T);
 20          s2=' ';
 21          s3=num2str(V);
 22          fprintf(fid,[s1 s2 s3 '\n']);
 23
 24          s1=num2str(1.01*D*T+n*T);
```

Figure 2.297

Click the Save icon to save the entered code. Then click the Run icon to run it (Figure 2.298).

Figure 2.298

Assume that we want to simulate the Buck converter of Example 33 with a switching frequency of 25 kHz and duty cycle of 60%. Run the MATLAB code with the inputs shown in Figure 2.299.

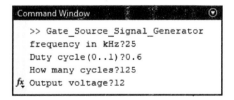

Figure 2.299

Open the schematic of Example 33 (Figure 2.300).

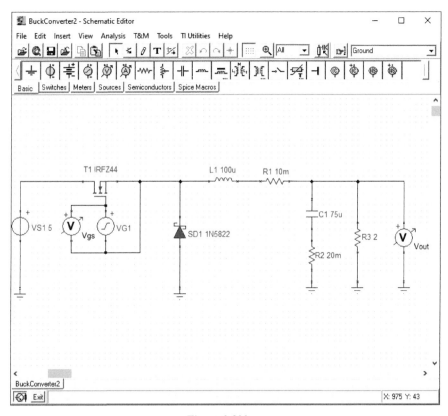

Figure 2.300

Double click the three dots behind the Signal box (Figure 2.301).

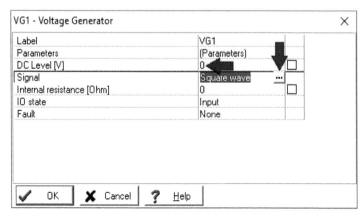

Figure 2.301

Click the Piece wise Linear Signal button (Figure 2.302).

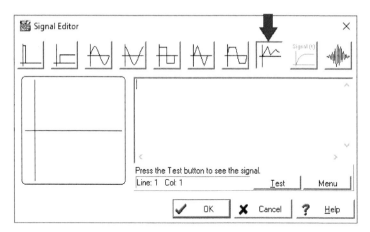

Figure 2.302

Click the Menu button. This opens a list for you. Click on the Open (Figure 2.303).

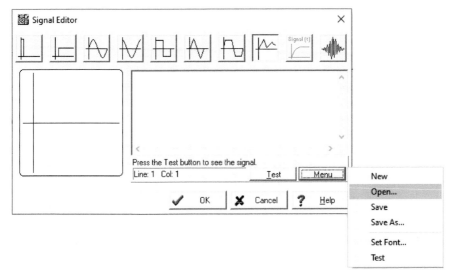

Figure 2.303

Go to the drive D and open the 'gate_source_signal.txt' (Figure 2.304). Then click the OK button in Figure 2.304.

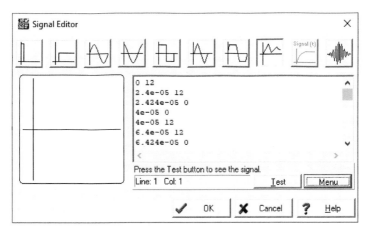

Figure 2.304

Run a transient analysis with the settings shown in Figure 2.305. The simulation result is shown in Figure 2.306.

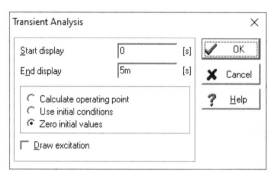

Figure 2.305

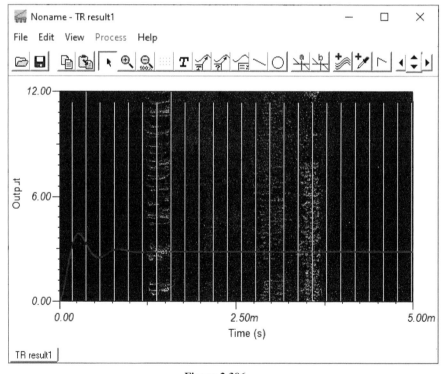

Figure 2.306

Remove the graph of output voltage (Vout) from the screen by clicking the View > Show/Hide curves and unchecking the V_{out} (Figure 2.307).

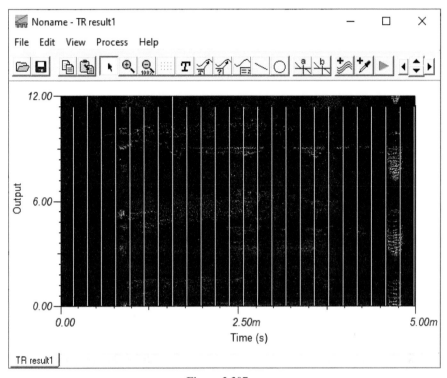

Figure 2.307

Use the magnifier icon to see the waveform better (Figure 2.308). According to Figure 2.309, the frequency of the pulse applied to the MOSFET is $\frac{1}{40.04\ \mu s} \approx 25$ kHz. According to Figure 2.310, the duty cycle of this pulse is $\frac{24.12\ \mu s}{40.04\ \mu s} \times 100\% = 60\%$. This shows that MATLAB code works correctly.

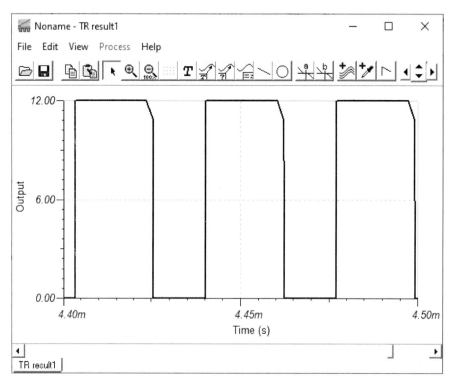

Figure 2.308

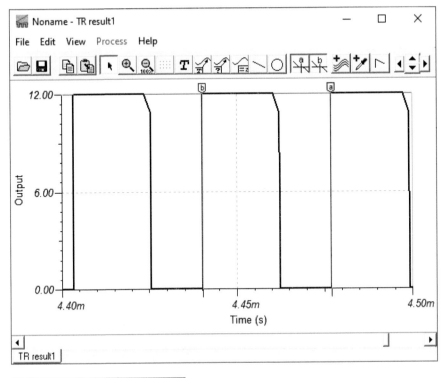

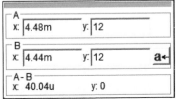

Figure 2.309

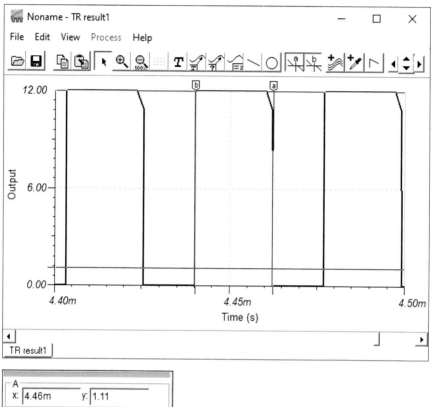

Figure 2.310

Click on the View > Show/Hide curves. Then check the V_{out} box and uncheck the V_{gs}. This shows the graph of output voltage on the screen (Figure 2.311).

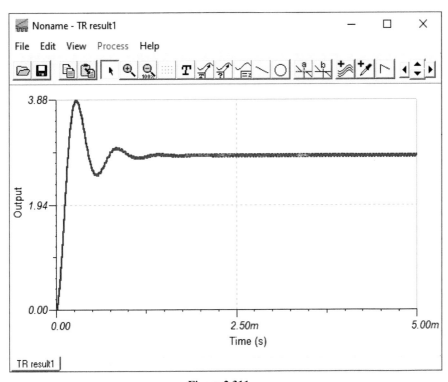

Figure 2.311

Zoom into the steady state region (Figure 2.312).

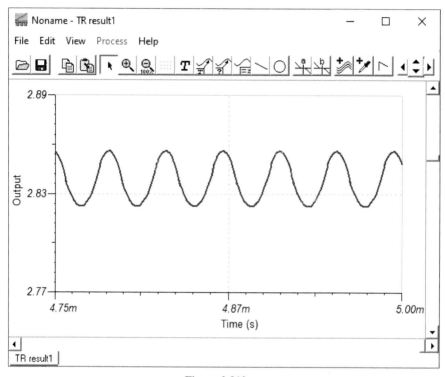

Figure 2.312

According to Figure 2.313, the output voltage ripple is 34.41 mV.

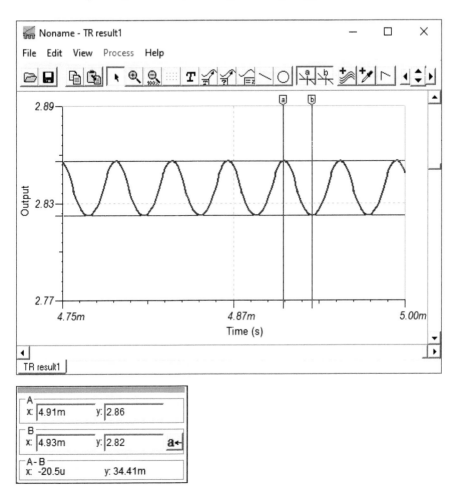

Figure 2.313

The average value of output voltage can be found by averaging the minimum and maximum of the output voltage. According to Figure 2.313, the minimum and maximum output voltage are 2.82 V and 2.86 V, respectively. Therefore, an average value of the output voltage is $\frac{2.82+2.86}{2} = 2.84\ V$. You can use the Process> Averages to measure the average value of the output voltage as well (Figure 2.314).

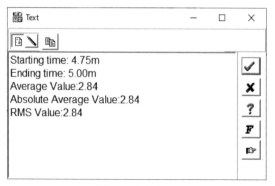

Figure 2.314

2.37 Exercises

1. Simulate the half-wave rectifier circuit with an RL load (Figure 2.315). Compare the result with the purely resistive load case.

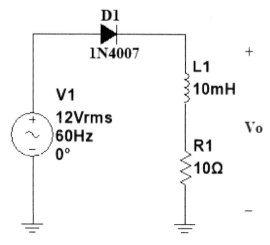

Figure 2.315

2. Figure 2.316 shows an op-amp clamp circuit with a non-zero reference clamping voltage. The clamping level is at precisely the reference voltage. Use TINA-TI to simulate the circuit and see the effect of the Reference Voltage source on the output.

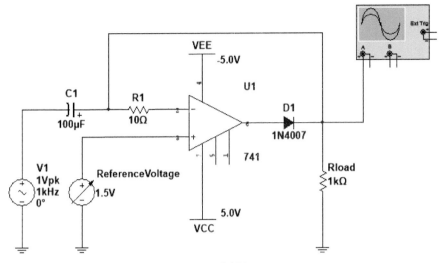

Figure 2.316

3. Measure the maximum output voltage swing for the circuit shown in Figure 2.317.

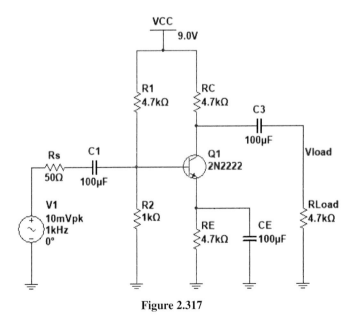

Figure 2.317

4. Assume the amplifier shown in Figure 2.318.

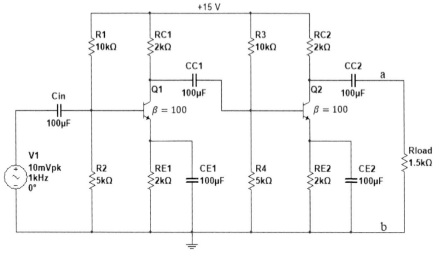

Figure 2.318

a) Use hand analysis to calculate the operating point of the circuit.
b) Use TINA-TI to verify results of part (a).
c) Use hand analysis to calculate the input impedance (impedance seen from source V1) and output impedance (impedance seen from points a and b) of the circuit.
d) Use TINA-TI to verify part (c).
e) Use hand analysis to calculate the overall gain ($\frac{V_{ab}}{V_1}$) of the circuit.
f) Use TINA-TI to verify part (e).

5. Assume that both transistors in Figure 2.318 are 2N2222. Use TINA-TI to draw the:

a) Input impedance.
b) Output impedance.
c) Frequency response of the amplifier.

References

[1] Razavi, B.: Fundamentals of microelectronics, 3rd edition, Wiley (2021)
[2] Rashid M.H.: Microelectronic circuits: Analysis and design, cengage learning (2016)
[3] Sedra, A., Smith, K., Carusone, T.C., and Gaudet, V.: Microelectronic circuits,8th edition, Oxford University Press, (2019)

Index

B

Buck converter 403, 417, 422, 424

C

Common mode rejection ratio 386
Coupled inductors 151, 155

F

Full wave rectifier 262, 267, 272

H

Half wave rectifier 211, 216,
227, 436

P

Phasor analysis 192

R

Rectifier 211, 216, 262, 267, 278
RLC Circuit 41, 66, 168,
185, 189

S

Simulation of electric circuits 1
Simulation of electronic
circuits 211
Step response of circuit 170

T

TINA-TI 1, 29, 60, 170, 224, 305
Total harmonic distortion 224, 317

V

Voltage gain 313, 314

About the Author

Farzin Asadi has a BSc in Electronics Engineering, MSc in Control Engineering, and PhD in Mechatronics Engineering. Currently, he is with the Department of Electrical and Electronics Engineering at Maltepe University, Istanbul, Turkey. Dr. Asadi has published more than 40 international papers and 15 books. He is also on the editorial board of 7 scientific journals. His research interests include switching converters, control theory, robust control of power electronics converters, and robotics.